re
i

83 Topics in Current Chemistry
Fortschritte der Chemischen Forschung

Biochemistry

Springer-Verlag
Berlin Heidelberg New York 1979

This series presents critical reviews of the present position and future trends in modern chemical research. It is addressed to all research and industrial chemists who wish to keep abreast of advances in their subject.

As a rule, contributions are specially commissioned. The editors and publishers will, however, always be pleased to receive suggestions and supplementary information. Papers are accepted for "Topics in Current Chemistry" in English.

ISBN 3-540-09312-5 Springer-Verlag Berlin Heidelberg New York
ISBN 0-387-09312-5 Springer-Verlag New York Heidelberg Berlin

Library of Congress Cataloging in Publication Data. Main entry under title: Biochemistry. (Topics in current chemistry ; v. 83) Bibliography: p. Includes index. CONTENTS: Deluca, H. F., Paaren, H. and Schnoes, H. K. Vitamin D and calcium metabolism. – Ullrich, V. Cytochrome P 450 and biological hydroxylation reactions. – Reden, J. and Dürckheimer, W. Aminoglycoside antibiotics. 1. Biological chemistry – Addresses, essays, lectures. 2. Vitamin D metabolism – Addresses, essays, lectures. 3. Cytochrome P 450 – Addresses, essays, lectures. 4. Aminoglycosides – Addresses, essays, lectures. I. Series.
QD1.F58 · vol. 83 · [QP509] · 540'.8s · [574.1'92] · 79-15156 ·

Typesetting and printing: Schwetzinger Verlagsdruckerei GmbH, 6830 Schwetzingen. Bookbinding: Konrad Triltsch, Graphischer Betrieb, 8700 Würzburg
2152/3140 – 543210

Contents

Vitamin D and Calcium Metabolism

Hector F. DeLuca, Herbert E. Paaren, and Heinrich K. Schnoes

Department of Biochemistry, College of Agricultural and Life Sciences, University of Wisconsin-Madison, 420 Henry Mall, Madison, Wisconsin, 53706, U. S. A.

Table of Contents

I Introduction

For a period between 1940 and 1968 the topic vitamin D chemistry remained a quiescent field, primarily because the work on the isolation and identification of the nutritional forms of vitamin D had been completed and a general feeling existed that the vitamin acted directly on the target tissues without metabolic alteration. This conclusion was supported by the early work on the metabolism of vitamin D in which Kodicek[1] concluded that although vitamin D was metabolized to demonstrable products, they appeared to be biologically inactive. With the development of chemical synthesis of radioactive vitamin D of high specific activity in the early 1960 era came the possibility of studying the metabolism and localization of truly physiologic doses of the vitamin[2,3]. Furthermore, rapid development in methods of chromatography for lipid soluble materials made possible the separation of vitamin D from its metabolites in a convenient and reproducible manner[4]. Finally, the development of mass spectrographic methods of structural determination of organic compounds and nuclear magnetic resonance spectrometers made possible identification of small amounts of important metabolites[5]. These developments coupled with the concept that vitamin D must be metabolically altered before it can function led to a demonstration of the first biologically active metabolites of vitamin D[6,7]. This resulted ultimately in the discovery that vitamin D is a precursor of at least one steroid hormone which is involved in the regulation of calcium and phosphorus metabolism[8]. This demonstration made clear the application of the active metabolites of vitamin D to the treatment of calcium, phosphorus and bone disorders[9]. Furthermore, it laid open the possibility that compounds could be synthesized as analogs of the active hormonal form of vitamin D and be used in a rather specific manner. These discoveries, therefore, reawakened the interest of the classical organic chemists in the problems of vitamin D chemistry. Ultimately this resulted in novel methods of synthesis of vitamin D and its metabolites and the synthesis of interesting and biologically active analogs. It is the purpose of this chapter to describe the development in our understanding of the metabolism of vitamin D, our latest understanding of its molecular mechanism of action at the target tissues and the resulting new advances in synthetic efforts on the vitamin D compounds and finally the description of important analogs of $1,25\text{-}(OH)_2D_3$ and their potential use biologically and medicinally.

II Historical

The existence of vitamin D was actually discovered as a consequence of human disease. Rickets has been known almost since antiquity and was quite clearly described in the literature around 1600. Glisson[10] and Whistler[11] are credited with the first clear description of the disease rickets which we now know is the childhood counterpart of vitamin D deficiency disease. With the industrial revolution which took place in the latter part of the 19th century came the changing of an agrarian society to an industrial society. The population in large measure spent much more time inside buildings which shielded them from ultraviolet light. In addition, the industrial plants produced large amounts of contaminating environmental pollutants which permitted only limited amounts of ultraviolet light to reach the population even when they were outdoors. Since vitamin D is not abundant in foods except for fish liver oils[12], a dietary source of vitamin D was also absent. In the low sunlight environments found in Northern Europe, North American and Northern Asia, we know in retrospect that insufficient amounts of ultraviolet light reached the skin of the population, thereby failing to produce necessary vitamin D and resulting in a deficiency disease. Early in this century therefore the disease rickets appeared in children in epidemic proportions in these parts of the world[13]. It was at this time that the concept of vitamins began evolving, starting with the work of Magendie[14] and Liebig[15] who attempted to maintain animals on a diet of chemically defined carbohydrate, fat, protein and minerals without success. These investigators concluded that there must be some vital substance necessary for life. In the United States agricultural chemists became interested in the fact that proximate analysis for carbohydrate, fat and protein failed to provide an adequate explanation of the adequacy or inadequacy of the diets of domestic animals[16]. This led the group at the University of Wisconsin to conclude that some micronutrients, heretofore unappreciated are required for the health and reproduction of domestic animals[16]. Beri-beri had appeared in high proportions in the prisoners of the Dutch East Indies and the physician Eichmann[17] had deduced that their diet of polished rice was responsible for the disease. Although he incorrectly concluded that the polishing of rice introduced a toxic substance, he nevertheless placed correct emphasis on diet as being responsible for the disease. His countryman, Grijns[18], was later able to show that the hulls of whole rice possessed a substance which could prevent or cure beri-beri demonstrating existence of an essential dietary factor. The Wisconsin group had begun to investigate the possibility that dietary components possessed micronutrients of considerable importance. Using rats McCollum[19] demonstrated the existence of a fat-soluble substance found in cod liver oil and butter fat which could support growth and prevent xeropthalmia in animals maintained on an otherwise chemically defined diet. Using the term coined by Funk[20] McCollum described this fat-soluble growth factor as vitamine A[21]. He similarly fed animals on the purified diets with cod liver oil as a source of vitamin A and discovered that the animals developed neurological symptoms. He was able to demonstrate that milk salts or lactose possessed a substance which would prevent this disease. He, therefore, called this substance a water soluble vitamin B[21]. Thus the vitamins became discovered in the 1910–1919 era. Undoubtedly, this discovery inspired Sir Edward Mellanby in Great Britian to consider that the disease rickets, then rampant in children of England, might be a dietary deficiency. Sir Edward Mellanby was able to produce a disease in dogs quite similar to the rachitic condition in children by feeding then a diet of oat meal and keeping them indoors[22]. By administering cod liver oil, Mellanby cured the disease. Since McCollum had shown the existence of fat-soluble vitamin A in cod liver oil, Sir Edward Mellanby concluded that vitamin A was responsible for curing the disease rickets. However, McCollum demonstrated quite clearly by bubbling oxygen through cod liver oil and heating the preparation that the growth promoting activity and antixeropthalmia activity could be destroyed but the ability to cure rickets remained[23]. He therefore concluded that the activity was the result of another fat-soluble vitamin which he called vitamin D. At the same time this work was being carried out, Viennese and British physicians were demonstrating that rickets in children could be cured by exposing them to ultraviolet light either of solar origin or artifical origin[24, 25]. Thus artificial light and cod liver oil were equal in their activity in curing this desease. Goldblatt and Soames[26] found that the livers taken from rachitic rats exposed to ultraviolet light possessed a material which cured rickets in other rachitic rats. This undoubtedly led Steenbock[27, 28] to demonstrate that ultraviolet light was able to activate a fat-soluble substance to become antirachitic and this substance could be found not only in skin but in a large variety of biological materials. He and his co-workers[29] as well as Hess and his co-workers[30] found the antirachitic material in the non-saponifiable fraction or the sterol fraction. This discovery led ultimately to the isolation and identification of vitamin D_2 from ergosterol irradiation mixtures[31, 32] and led to the elimination of rickets as a major medical problem since irradiation of foods could be used to induce vitamin D activity in them[33]. In 1937 Windaus and his collaborators synthesized provitamin D_3, namely 7-dehydrocholesterol, and subsequently isolated and identified vitamin D_3 from irradiation mixtures of the synthetic material[34, 35]. This essentially ended the isolation and identification work on the dietary D vitamins and a quiescence settled over the study of vitamin D chemistry. However, considerable work was expended on the mechanism of action even while the isolation and identification of the D vitamins was in progress. Howland, Kramer and Shipley[36, 37] provided the first evidence that the major failure to calcify bone in rickets is an insufficient supply of calcium and phosphorus to the mineralization sites. Thus a diagnostic measure of vitamin D deficiency was a low product of calcium concentration times phosphorus concentration in blood[36]. Orr[38] was the first to demonstrate that vitamin D improved intestinal absorption of calcium, a discovery which was not considered popular at the time. Nicolaysen[39, 40] in the 1935–1943 era was able to show conclusively that the vitamin D directly stimulates intestinal calcium absorption and furthermore was able to demonstrate that the ability of the intestine to adapt to dietary calcium levels required the presence of vitamin D[41]. These basic discoveries have been confirmed many times by a variety of investigators using a variety of techniques[42] and remains as one of the basic functions of the D vitamins. In 1952 Carlsson[43], provided the first true insight in the mechanism of action of vitamin D on bone. Rather than acting directly on the mineralization process, Carlsson demonstrated that vitamin D was responsible for the mobilization of calcium from previously formed bone. This important discovery was confirmed by Nicolaysen and Eeg-Larsen[44]. It was later shown that this mechanism involves not only vitamin D but also the parathyroid hormone[45]. Another site of vitamin D action was discovered by Harrison and Harrison[46], who demonstrated that vitamin D activates the transport of phosphate across the intestinal membrane. This process has now been shown to be quite independent of the calcium transport process although its mechanism remains unknown.

In 1963 radioactive vitamin D of high specific activity was successfully synthesized chemically[2, 3]. This preparation was labelled in the 1 and 2 positions and provided the first material whereby truly physiologic amounts of vitamin D could be studied from a metabolic point of view. This led to a demonstration of the existence of polar metabolites of vitamin D possessing biological activity equal to or better than the parent vitamin[6, 7]. Furthermore, the polar metabolites of vitamin D could stimulate intestinal calcium absorption and mobilization of calcium from bone more rapidly than vitamin D itself[47]. Thus the existence of potent metabolites of vitamin D became known in 1967. By 1968 the first of the potent forms of vitamin D was isolated in pure form and identified as 25-hydroxyvitamin D_3 (25-OH-D_3)[48]. Its structure was quickly confirmed by chemical synthesis[49] which led to the preparation of radioactive 25-OH-D_3[50]. Using this preparation it was then possible to show that it, like its vitamin D precursor was rapidly metabolized further and appeared in the target tissues of intestine and bone before those tissues responded by increasing intestinal calcium absorption and the mobilization of calcium from bone[51, 52]. In 1971 the most potent and hormonal form of vitamin D was isolated in pure form from chick small intestine and identified as 1,25-dihydroxyvitamin D_3 (1,25-$(OH)_2D_3$)[53, 54]. Its structure was then quickly confirmed by chemical synthesis proving its structure to be 1α25-dihydroxyvitamin D_3 (1α,25-$(OH)_2D_3$)[55]. This conclusion was finally supported by the synthesis of the epimer 1β,25-dihydroxyvitamin D_3 (1β,25-$(OH)_2D_3$) and the demonstration that the natural product was identical with 1α,25-$(OH)_2D_3$[56].

Our understanding of the metabolism of vitamin D is yet to be completed. Additional metabolites have been isolated and identified although their biological function remains unknown. In addition, the mechanism whereby 1,25-$(OH)_2D_3$ stimulates intestinal calcium transport and intestinal phosphate transport, the mobilization of calcium from bone and the mineralization of bone remain to be elucidated. This chapter, therefore, is a progress report and more information can be expected in regard to the metabolism of vitamin D, its regulation and its mechanism of action. Furthermore, important and interesting analogs of the metabolites of vitamin D can be expected which will have special biological properties. Thus, the topic of vitamin D chemistry is certainly current.

III Metabolism of Vitamin D_3 (Fig. 1)

The clear demonstration that physiologic amounts of vitamin D_3 are metabolized in large proportions to metabolites possessing biological activity equal to or greater than vitamin D_3 itself came following completion of chemical synthesis of [1,2-^{3}H]-vitamin D_3 of high specific activity[2, 3]. The major circulating metabolite of vitamin D_3 was then isolated from plasma of pigs and identified chemically as 25-OH-D_3[48]. The liver was shown to be the site of this conversion[57, 58] although evidence has been presented that some 25-hydroxylation occurs in kidney and intestine of chicks[59]. A reexamination of this has revealed that certainly in the rat and probably in the chick, the liver is the major, if not sole, site of 25-hydroxylation[60, 61]. Even during the isolation of 25-OH-D_3 prior to its identification, the existence of metabolites more polar than 25-OH-D_3 was noted[48], but they were not

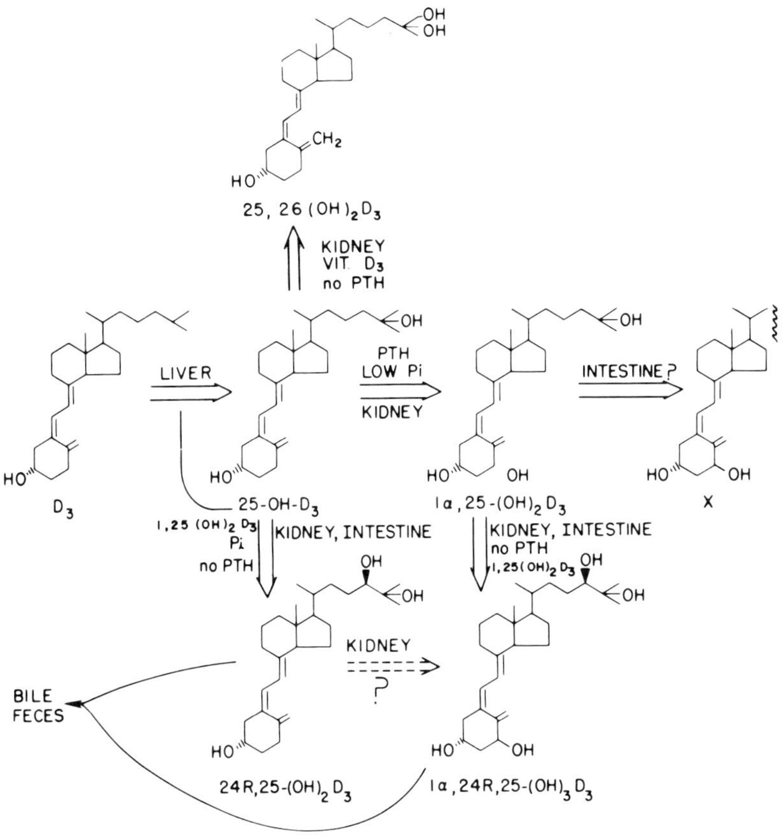

Fig. 1. The metabolism of vitamin D: 1978

further characterized or their biological activity examined. Haussler, Myrtle and Norman detected a metabolite of vitamin D_3 more polar than 25-OH-D_3 in the nuclear fraction of small intestine[62]. At the same time, Lawson, Wilson and Kodicek also located this metabolite of vitamin D and found that during its biosynthesis the 1α-^{3}H from $1\alpha^3$H-vitamin D_3 was lost[63]. They used this observation to suggest that this metabolite possessed a modification on carbon 1, although other explanations were clearly possible. Independently, Ponchon *et al.* studied extensively the metabolites of vitamin D in a variety of tissues in both rats and chicks, demonstrating the existence of that was termed peak V as well as other more polar metabolites[64].

Synthesis of radioactive 25-OH-D_3[50] allowed a pursuit of its metabolites and it was soon demonstrated that the polar metabolite of the small intestine, designated as peak V in the Wisconsin laboratories, peak p in the Cambridge laboratories, and 4B in the Riverside laboratories, was formed from 25-OH-D_3[51]. Although Myrtle, Haussler and Norman[62] first reported marked biological activity of the intestinal metabolite, this finding could not be uniformly reproduced[64]. Haussler et al.[65] discovered that the intestinal metabolite initiated intestinal calcium transport very rapidly and that its activity decreased very rapidly, accounting for the divergency of biological activity reports. This very rapid activity was confirmed in two laboratories[66, 67] and led to an intensive approach to the isolation and identification.

In early 1971 the isolation of this metabolite from a target tissue (the intestines) of 1500 rachitic chickens given a single dose of radioactive vitamin D_3 was accomplished[53, 54]. Isolation involved approximately ten chromatographic procedures through which the lipid extract of the intestines was processed. This, however, led to a metabolite fraction contaminated with another trihydroxy sterol which could then be separated upon derivatization as the 25-hydroxy-trimethyl silyl ether and subsequent chromatography on Sephadex LH-20. The isolation of the metabolite from the intestine involved primarily the use of liquid-gel partition chromatography which was introduced as a major chromatographic technique for the study of vitamin D metabolism. From the 1500 rachitic chickens given the radioactive vitamin D_3, two micrograms of 25-trimethylsilyl ether derivative was isolated in pure form and the structure of the metabolite unequivocally identified by chemical means as 1,25-$(OH)_2D_3$. Proof that this compound was 1α,25-$(OH)_2D_3$ was provided by chemical synthesis of it[55] and its 1β isomer[56]. While this isolation was beeing carried out, Fraser and Kodicek[66] discovered that the site of production of the intestinal metabolite was kidney and were able to produce the metabolite *in vitro* by incubation of chick kidney homogenates and isolated mitochondria with 25-OH-$[^3H]D_3$. From *in vitro* incubations, Lawson et al.[67] were able to obtain several micrograms of metabolite in approximately 30% purity. From this preparation some evidence for the 1,25-$(OH)_2D_3$ structure was obtained. It, therefore, became clear that 25-OH-D_3, which was produced predominantly if not exclusively in the liver, is hydroxylated to 1,25-$(OH)_2D_3$ in the kidney. The exclusive synthesis of 1,25-$(OH)_2D_3$ in the kidney could easily be demonstrated in nephrectomized rats given radioactive 25-OH-D_3 intravenously[68]. These animals failed to produce 1,25-$(OH)_2D_3$, while sham-operated and uremic controls were clearly able to perform the conversion. The ability of chick kidney homogenates and mitochondria to carry out the conversion of 25-OH-D_3 to 1,25-$(OH)_2D_3$ was confirmed in two different laboratories[68, 69].

Using the nephrectomized animal it was possible to demonstrate quite clearly that 1,25-$(OH)_2D_3$ is the active form of vitamin D_3 in the initiation of intestinal calcium transport[70, 71], bone calcium mobilization[72] and intestinal phosphate transport[73]. Briefly, nephrectomized animals do not show these responses to 25-OH-D_3 when given in physiologic amounts whereas 1,25-$(OH)_2D_3$ produces these responses whether the kidneys are present or not. Thus it is quite evident that 25-OH-D_3 and vitamin D_3 itself cannot be considered metabolically active forms at physiologic concentrations. However, it is well known that vitamin D or 25-OH-D_3 can produce responses in nephrectomized animals when given in large amounts[74, 75]. In addition in isolated cultures of bone[76, 77], in vascularly perfused intestine[78] and in isolated intestinal cultures[79, 80], it can be shown that large amounts of 25-OH-D_3 can cause the mobilization of calcium from bone, the transport of calcium across intestinal membrane and the production of calcium binding protein. Exactly what the reason is for responses to pharmacological amounts of these compounds is not known. Two possibilities are now considered: one, that the receptor molecules or systems will respond to the unhydroxylated forms of vitamin D if present in large concentrations[81, 82]; or alternatively, small amounts of *trans* vitamin D or isotachysterol contaminants, which have their 3-hydroxyl in a pseudo 1 position, might produce a 1,25-$(OH)_2D_3$- like response[83–85]. In any case, from physiologic point of view it is quite clear that the 25-OH-D_3 to 1,25-$(OH)_2D_3$ pathway is an obligatory one for physiologic amounts of vitamin D to carry out its known functional role.

Following the demonstration that 1,25-$(OH)_2D_3$ or a further metabolite is the final metabolically active form of the vitamin, work has been carried out with radioactive 1,25-$(OH)_2D_3$ in an effort to learn whether it is further metabolized[86–88]. At the time the intestine and bone respond to this compound, no other metabolites have yet been demonstrated, leaving the impression that 1,25-$(OH)_2D_3$ is not further metabolized before it functions. However, experiments with 25-OH-[26,27-^{14}C]D_3 have shown that 7% of this compound is metabolized to $^{14}CO_2$ in rats[89, 90] and chicks[91]. Nephrectomy prevents this metabolism and 1,25-$(OH)_2$[26,27-^{14}C]D_3 is metabolized to the extent of 30% in both normal and anephric rats to $^{14}CO_2$[90]. Thus, 1,25-$(OH)_2D_3$ is converted to a compound lacking a portion of the side chain. This reaction occurs rapidly enough (within 3 hours after injection) to be of functional significance. Recently a metabolite of 1,25-$(OH)_2D_3$ without the 26 and 27 carbons has been isolated in pure form and identified as 23-carboxy 23-27-tetranor 1α-hydroxyvitamin D_3 (R. Esvelt, H. Schnoes and H. DeLuca, unpublished results). Its biological activity has not yet been studied, however. Nevertheless, the existence of the side-chain cleavage metabolism of 1,25-$(OH)_2D_3$ has been clearly established, but its physiological significance remains unknown.

During the course of the isolation and identification of 1,25-$(OH)_2D_3$, other metabolites of vitamin D_3 were detected, isolated and identified. In particular from the peak V region described by Blunt and DeLuca, two metabolites were isolated and tentatively identified. The first, which was termed peak Va, was initially identified as 21,25-dihydroxyvitamin D_3[92], but upon more thorough characterization, the correct structure was deduced to be 24,25-$(OH)_2D_3$[54]. This compound has proved to be a major metabolite of vitamin D_3 whose exact function is still under examination. Chemical synthesis of this compound has been completed[93] and the two possible stereoisomers prepared independently by Ikekawa and his group[94] und by the Hoffmann-LaRoche group[95]. The natural form of the 24,25-$(OH)_2D_3$ has proved to be 24R,25-$(OH)_2D_3$[96]. This has been deduced by cochromatography of the natural radioactive product with the synthetic isomers as their silyl ether derivatives. The natural product cochromatographs exactly with the 24R compound with no radioactivity appearing with the 24S compound. In the rat, the 24R compound has biological activity in intestinal calcium transport, calcification of bone and mobilization of calcium from bone approaching that observed with 25-OH-D_3 itself[96]. Furthermore, the administration of radioactive 24,25-$(OH)_2D_3$ to rats reveals that it is rapidly excreted.

On the other hand, 24,25-$(OH)_2D_3$ has very low activity in the chicken[97–99] and is rapidly eliminated. So fast is the elimination that little 1,24,25-$(OH)_3D_3$ can be detected in this species[100]. It is unknown if 24,25-$(OH)_2D_3$ is rapidly excreted in mammals receiving sufficient amounts of vitamin D or not. Nevertheless, in the known biochemical functions of the vitamin, 24,25-$(OH)_2D_3$ is relatively inactive in anephric animals[93, 101], suggesting the necessity for 1-hydroxylation. This led to the isolation and identification of 1,24,25-$(OH)_3D_3$[102]. The configuration of the 24 hydroxyl in the naturally produced form of this metabolite is also R[103]. This metabolite is more potent than 24,25-$(OH)_2D_3$ but less potent than 1,25-$(OH)_2$-D_3[103, 104], and has weak activity in the chicken[97, 98]. It is about one order of magnitude less active than 1,25-$(OH)_2D_3$ in receptor binding and other *in vitro* tests such as bone resorption in culture. A thorough study of its biological activity in the rat has revealed that it shows no advantage over 1,25-$(OH)_2D_3$ in intestinal calcium and phosphorus transport, mineralization of rachitic cartilage and bone mineral mobilization, and is less active than 1,25-$(OH)_2D_3$ in all of these systems[104]. It is, therefore, clear that 24R-hydroxylation of vitamin D metabolites does not improve, but actually diminishes, biological activity in the systems known to be vitamin D dependent.

There have been reports that 24,25-$(OH)_2D_3$ is a mineralization form of the vitamin since it restores "calcification front" in osteoid tissue much better than does a lower dose of 1,25-$(OH)_2D_3$ and thus is more effective against osteomalacia[105]. Henry et al.[106] have proposed that 24,25-$(OH)_2D_3$, together with 1,25-$(OH)_2D_3$, functions in the suppression of parathyroid gland size and secretion, a result supported to some degree by the work of Canterbury et al.[107]. Garabedian et al.[108] report that 24,25-$(OH)_2D_3$ is particularly effective in stimulating cartilage tissue growth in culture. Henry and Norman[109] have reported that 24,25-$(OH)_2D_3$ is essential for normal embryonic development in chicks. These interesting studies may suggest previously unknown functions of vitamin D satisfied preferentially by 24R hydroxylated forms of vitamin D. It is in this manner that the 24R hydroxy group may participate in the functions of vitamin D. However, it must be noted that the above suggested functions are far from established and require more definitive experiments *in vivo* before they can be accepted.

A suggestion by Kanis et al.[110] that 24,25-$(OH)_2D_3$ functions directly in man to facilitate absorption of calcium cannot be verified and is likely the result of technical problems associated with interpretation of clinical data.

When the 24-hydroxylated vitamin D metabolites were being investigated, the 24 R and 24S isomers of 24-OH-D and 24,25-$(OH)_2D$ were synthesized[111, 112, 94, 95]. Testing of the biological activity revealed the 24S hydroxy compounds to be less active than the corresponding 24R hydroxy compounds[96, 112]. 24R-OH-[24S-^{3}H]D_3 and 24S-OH-[24R-^{3}H]D_3 were prepared and their metabolism studied[113]. The results demonstrated that 25-OH-D_3-1-hydroxylase discriminates against the 24S hydroxyl group. Thus the 24S compound is converted to 24S,25-$(OH)_2D_3$ but is not further activated to the corresponding 1,24,25-$(OH)_3D_3$, while the 24R compound is. 1,24S,25-$(OH)_3D_3$ and 1,24R,25-$(OH)_3D_3$ are equally biologically active[104] confirming this deduction. The discrimination against the 24S hydroxyl configuration in 24,25-$(OH)_2D_3$ also argues for the functional significance of 24-hydroxylation as does the fact that its production is regulated (see section on regulation). However, at this point, the functional significance of the 24R hydroxyl group is far from established and remains to be investigated.

Because of the substitution on the 24 carbon and because chicks and fowl discriminate against vitamin D compounds which have a methyl group on the 24 carbon[114], it is indeed possible that a 24 substitution is a signal in the chick or other birds for elimination. Thus the 24-hydroxylations in these species may represent the first reaction in the metabolic elimination of potentially toxic vitamin D compounds[97]. It is possible, therefore, that the birds discriminate against the vitamin D_2 compounds simply because they are analogous to the natural excretion product (24R,25-$(OH)_2D_3$) of the natural form of vitamin D, namely vitamin D_3[97]. In agreement with this idea, it has already been demonstrated that vitamin D_2 and its metabolites are rapidly metabolized and excreted via the bile into the feces[115].

25,26-Dihydroxyvitamin D_3 (25,26-$(OH)_2D_3$) has also been isolated, identified[116] and chemically synthesized[117, 118]. This compound has only weak biological activity in stimulating intestinal calcium transport and has little or no activity in the other systems known to be responsive to the vitamin. Its intestinal calcium transport activity requires kidney function presumably to convert it to 1,25,26-$(OH)_3D_3$[117]. Chemical synthesis of the two possible isomers has been completed[118] and with these, the natural configuration has been shown to be R[119]. Its appearance is quantitatively smaller than that of 24,25-$(OH)_2D_3$ and its role, if any, is not yet known.

Throughout the course of work on the vitamin D_3 metabolism, parallel work has occasionally been carried out with the vitamin D_2 series. 25-OH-D_2 has also been isolated and chemically identified[120]. Furthermore, 1,25-$(OH)_2D_2$ has been prepared *in vitro* with chick kidney mitochondria and its structure unequivocally elucidated[121]. The biological activity of the vitamin D_2 compounds has been assessed in the rat and found to be identical in every respect to the vitamin D_3 compounds. In the bird, however, as described above, the vitamin D_2 compounds, including 1,25-$(OH)_2D_2$, are approximately 1/5 to 1/10 as active as their vitamin D_3 counterparts[114]. Of some interest is that Upjohn chemists have successfully synthesized the 25-OH-D_2 and the 25-OH-24-epiD_2. The 25-OH-24-epiD_2 is much less active than 25-OH-D_2 (J. A. Campbell, L. Reeve and H. F. DeLuca, unpublished results).

New world monkeys also discriminate against vitamin D_2 in terms of their biological response[122]. Although the mechanism of this discrimination has not yet been determined, it might well be related to the mechanism described for the chick. In this regard it should be noted that the chick possesses the enzymes necessary for the synthesis of the active forms of vitamin D_2[123]. That is, the 25-hydroxylase of the liver of chicks does not discriminate against the vitamin D_2 molecule. Furthermore, the 1-hydroxylase system and the 24-hydroxylase system of the chick kidney do not discriminate against 25-OH-D_2[123]. However, following the injection of radioactive vitamin D_2 little 25-OH-D_2 is found in the circulation[124] and little 1,25-$(OH)_2D_2$ is found in the intestine[123]. Since the intestine of the chick also responds poorly to 1,25-$(OH)_2D_2$ as compared to 1,25-$(OH)_2D_3$, it seems that the bird discriminates against all forms of vitamin D with the substitution on the 24 carbon and that the mechanism of discrimination is likely to be one in which 24 substituted vitamin D's are rapidly metabolized and excreted[114]. An intestinal cytosol binding protein for 1,25-$(OH)_2D_3$ in the chick, which is considered a receptor, binds the 1,25-$(OH)_2D_2$ just as well as it binds the 1,25-$(OH)_2D_3$[81, 82, 125].

In addition, in chick embryonic intestine, 1,25-$(OH)_2D_2$ and 1,25-$(OH)_2D_3$ are equally effective in stimulating production of calcium binding protein (O. Parkes and H. F. DeLuca, unpublished results). These results provide evidence that the discrimination does not take place at the receptor level at least.

Little is known regarding the further metabolism of the vitamin D metabolites to excretion products. Vitamin D sulfates have been described[126, 127], but have not yet been adequately demonstrated to be a part of the normal metabolic scheme or to be a normal excretion product. The primary route of excretion appears to be bile with less than 4% of the total radioactivity of vitamin D_3 appearing in the urine[2, 128, 129]. No biliary metabolites have yet been identified, although claims of glucuronides or sulfates have appeared but without adequate chemical identification[130, 131].

Vitamin D sulfate has been reported to be present in milk[132–134] but the identification of it is far from convincing. Nevertheless, the claim is made that this water soluble form of vitamin D is more active in curing rickets in children than is ordinary vitamin D, a claim made with little or no scientific support.

IV The Biogenesis of Vitamin D

Vitamin D is not widely distributed in nature being quite low in the plant world except special circumstances[12, 135]. It is found in sizable amounts in livers of cartilaginous aquatic animals such as sharks and in fish[330]. The source of vitamin D in these species is still unknown. There is evidence that it arises by means of a non-

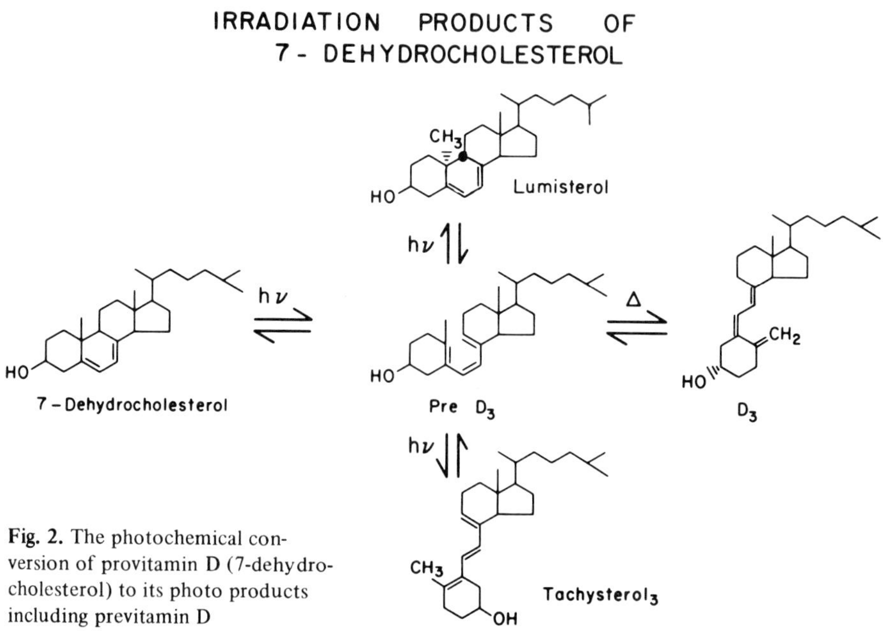

Fig. 2. The photochemical conversion of provitamin D (7-dehydrocholesterol) to its photo products including previtamin D

photochemical process[33, 136], but the evidence is not convincing. Other suggestions include accumulation of vitamin D from plankton or other small organisms from the sea surface since vitamin D may not be metabolized in fish and sharks. Certainly this represents an unsolved biological problem of considerable interest.

Vitamin D is produced in skin as a result of ultraviolet irradiation, a phenomenon realized in 1919–1924[24, 25]. It is known that ultraviolet light brings about a photolysis of 5,7-diene sterols giving rise to previtamin D and other photoisomers, lumisterol and tachysterol[137] (Fig. 2). The previtamin D spontaneously isomerizes to vitamin D to yield an equilibrium mixture favoring vitamin D. This photolysis known to occur in organic solvents also occurs in epidermis of skin. Ultraviolet light is known to penetrate to the site of 7-dehydrocholesterol in epidermis[138]. Recently, the antirachitic substance produced in rat skin by ultraviolet irradiation has been isolated in pure form and unequivocally identified as vitamin D_3[139, 140]. The kinetics of its production in skin is identical to that in organic solvents suggesting that proteins or enzymes are not involved. This initial step in vitmain D biochemistry therefore does not appear regulated.

V Enzymes Which Carry Out the Metabolism of Vitamin D_3 (see Fig. 1)

1 Vitamin D_3-25-Hydroxylase

There is little doubt that the liver is the major site of 25-hydroxylation of vitamin D_3[57, 58, 141]. There is some question whether this conversion occurs to any significant degree in tissues other than the liver. Tucker et al.[59] have reported that chick kidney and intestine possess 25-hydroxylating ability. The conversion in chick but not rat intestine has been confirmed in our laboratory[60, 61], but so far confirmation in the kidney has not been possible. Hepatectomy markedly reduces the ability of rats to produce 25-OH-D_3[58, 142]. Although some of this metabolite has been detected in hepatectomized rats, the amount is very small in the face of very high circulating concentrations of vitamin D_3[142]. Ordinarily vitamin D_3 is rapidly taken up in large amounts in the liver and does not circulate in high concentrations. In fact, it circulates in man at a concentration of 1 ng/ml[143] (Shepard and DeLuca unpublished). Thus the contribution of extra-hepatic tissue to 25-hydroxylation is likely small and perhaps absent under physiologic circumstances. In the bird, this hydroxylation may occur in part in the intestine especially following oral administration[60, 61].

Liver homogenates carry out the 25-hydroxylation when supported by glucose-6-phosphate and magnesium ions utilizing the endogenous glucose-6-phosphate dehydrogenase[141, 144]. However, the activity can be increased by the addition of an NADPH generating system[144]. Cell fractionation has revealed that two cell components are necessary for maximal rates of hydroxylation; the microsomal fraction and the cytosolic fraction[144]. The cytosol contains a heat labile factor which both protects the substrate and stimulates conversion to 25-OH-D_3. This might well be analogous to the systems described for cholesterol biosynthesis and for bile acid biosynthesis[145, 146]. A recent report that 25-hydroxylation of vitamin D occurs in mitochondria is likely the result of an impure mitochondrial preparation[147]. The K_m for vitamin D_3 is 4 x 10^{-8}M for both homogenates and microsomes and all the 25-hydroxylation in hepatic homogenates can be accounted for by microsomal activity (T. Madhok and H. F. DeLuca, unpublished results). The microsomal system will also hydroxylate vitamin D_2 and dihydrotachysterol$_3$.

The microsomal 25-hydroxylation of vitamin D_3 is markedly suppressed by the prior administration of vitamin D[148]. The degree of this suppression and the length of time of the suppression is dependent upon the dose of the vitamin D given. The suppression appears to be related to the hepatic level of 25-OH-D_3. External 25-OH-D_3 does not enter the liver in appreciable amounts and hence does not suppress the 25-hydroxylase[149]. It is unknown whether this suppression is mere product inhibition by 25-OH-D_3 or whether the 25-OH-D_3 alters the 25-hydroxylase, rendering it inactive. The administration of large amounts of vitamin D_3 will overcome this suppressed 25-hydroxylation and it is uncertain whether this is a competition with the 25-OH-D_3 of the liver releasing the microsomal 25-hydroxylation or whether it is due to extra-microsomal 25-hydroxylation which occurs at large substrate concentrations. In any case, the suppression can be overcome with increasing amounts of vitamin D[60, 150]. Thus the regulation phenomenon is observed only at physiologic doses of the vitamin and large amounts of vitamin D can increase the circulating levels of 25-OH-D_3[48, 151, 152]. The 25-hydroxylase is not influenced by dietary calcium, serum calcium, serum phosphorus or parathyroid hormone (T. Madhok and H. F. DeLuca, unpublished results).

Vitamin D-25-hydroxylase of rat liver is a mixed function monooxygenase as demonstrated by the incorporation of $^{18}O_2$ into the 25 position of vitamin D[153]. The reaction is inhibited by metyrapone, and CO:O_2 of 9:1. The CO inhibition is released with white light. Vitamin D inhibits aminopyrine demethylase. Thus all evidence currently available suggests this to be a cytochrome P-450 system of the type shown in Fig. 3. Phenobarbital does not induce the vitamin D-25-hydroxylase despite claims to the contrary (T. Madhok and H. F. DeLuca, unpublished results)[153, 154].

There has been, however, work carried out regarding the possible interference with vitamin D metabolism by the administration of anti-epileptic drugs such as

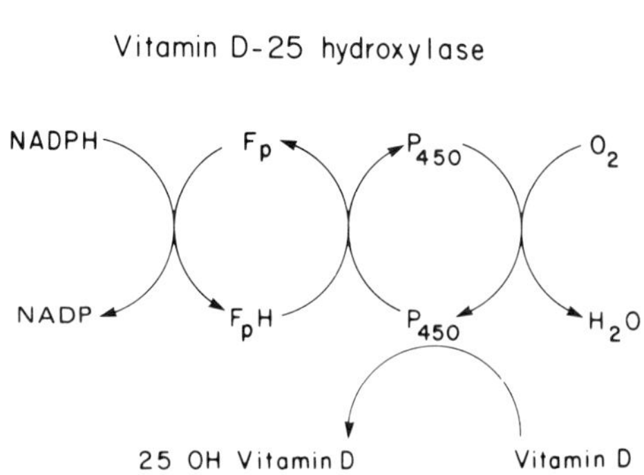

Fig. 3. The expected mechanisms of vitamin D 25-hydroxylation in liver tissue

phenobarbital and dilantin. Patients on longterm therapy show reduced circulating levels of 25-OH-D_3 and evidence of bone disease[153, 154]. There is considerable disagreement regarding the extent of the drug-induced osteomalacia, but there is no doubt that it does appear. Hahn and collaborators have suggested that phenobarbital and dilantin cause a rapid degradation of vitamin D and its metabolites to inactive products[158]. This interesting idea has not yet received sufficient biochemical support to warrant its acceptance. Alternatively, it has been suggested that phenobarbital and dilantin treatment interfere with 25-hydroxylation[159]. Thus the metabolism of vitamin D under conditions of drug-induction is one which merits further investigation.

2 25-Hydroxyvitamin D_3-1α-Hydroxylase and 25-Hydroxyvitamin D-24R-Hydroxylase

Perhaps the most important hydroxylation reaction from a physiologic point of view is that which occurs in the kidney. The 25-OH-D_3-1α-hydroxylase occurs exclusively in the kidney[66, 68] and it has been shown to be exclusively mitochondrial in the bird[66, 68, 160]. In mammals the system is found in the microsomes (Y. Tanaka and H. F. DeLuca, unpublished results). Although there is some evidence that the 1-hydroxylase of chick kidney is located in the innermitochondrial membrane, this has not yet been firmly established. With $^{18}O_2$ experiments it has been demonstrated that the 1-hydroxylase is a mixed-function oxidase using molecular oxygen for incorporation into the substrate 25-OH-D_3[161]. The hydroxylation in intact mitochondria is supported by succinate or malate or almost any Krebs cycle substrate[160, 162]. In intact mitochondria this hydroxylation requires the electron transport chain and oxidative phosphorylation for maximal activity[162]. Antimycin A blocks the reaction completely when supported by these substrates, whereas cyanide, rotenone and dinitrophenol interfere with the reaction to the extent of 50%[160]. Although a report has appeared that antimycin A does not block the renal 1-hydroxylase[163], great care must be taken in interpretation of these results since antimycin A is relatively unstable and if it fails to inhibit, its effectiveness in blocking the respiratory chain must be demonstrated. When the mitochondria are swollen with calcium ions, NADPH can be used for reducing equivalents[162]. NADH is not used, illustrating that the correct electron donor in NADPH. The 1-hydroxylase is blocked by carbon monoxide-oxygen mixtures as originally demonstrated by Fraser and Kodicek[66] and subsequently in our laboratory[162]. This inhibition is alleviated by white light[162] and more recently it has been shown to be alleviated more specifically by light of 450 nm wavelength[163]. Furthermore, glutethimide and metyrapone both inhibit the reaction which suggests that indeed the reaction is dependent upon a cytochrome P-450[162]. The reaction, therefore, appears analogous to the steroidogenesis systems of the adrenal gland. The existence of a cytochrome P-450 in the isolated chick mitochondria which is reducible by malate was clearly demonstrated by spectral means[114]. Since this spectrum only appeared when malate or other mitochondrial substrates were added, it is clear that the P-450 spectrum is of mitochondrial origin and not due to possible microsomal contamination. This finding was followed by solubilization of the chick kidney 1α-hydroxylase

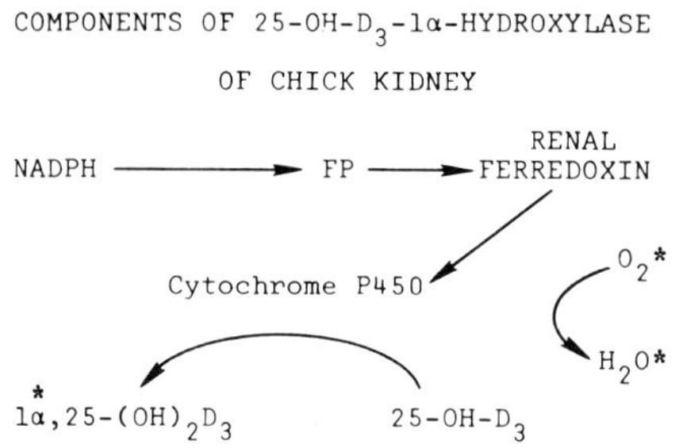

Fig. 4. The mechanism of 1-hydroxylation of 25-OH-D_3 in chick renal mitochondria. This system has been proved by the isolation of all three components and its subsequent reconstitution

system and the isolation of a P-450 fraction[164]. This P-450 fraction when recombined with beef adrenal ferredoxin and beef adrenal ferredoxin reductase together with NADPH carries out the hydroxylation of 25-OH-D_3 to 1,25-$(OH)_2D_3$[164]. Recently the P-450 preparation has been considerably improved and the chick renal ferredoxin has been isolated in almost pure form[165]. When these are recombined with beef adrenal flavoprotein and NADPH, the hydroxylation occurs to a very large extent with a single product appearing, namely 1,25-$(OH)_2D_3$. These experiments demonstrate clearly that the 1-hydroxylation is a P-450 dependent reaction whose mechanism involves the sequential reduction of flavoprotein, renal ferredoxin and cytochrome P-450 (Fig. 4). The K_m for 25-OH-D_3 is 1-2.2 x 10^{-6}M and the enzyme does not appear to act directly on vitamin D_3 or 24-OH-D_3[160], demonstrating the importance of the 25-hydroxyl group for interaction with the enzyme.

25-OH-D_3-24-hydroxylase has also been demonstrated to occur in the kidney[101, 166] but not exclusively so[167]. This enzyme also hydroxylates 1,25-$(OH)_2D_3$ to form 1.24R,25-$(OH)_3D_3$[103]. A 24-hydroxylase has also been located in intestine which apparently is more specific for 1,25-$(OH)_2D_3$[167]. Another 24-hydroxylase has been reported in cartilage tissue[168]. The chick kidney reaction occurs in the mitochondria and in general is similar to the 25-OH-D_3-1α-hydroxylase[166]. In intact mitochondria it is supported by Krebs cycle substrates, utilizes molecular oxygen and is blocked by inhibitors of oxidative phosphorylation. In calcium swollen mitochondria, NADPH will support the reaction and under these circumstances the reaction is independent of the electron transport system and oxidation phosphorylation. However, this reaction is not carbon monoxide sensitive and does not appear to be dependent upon cytochrome P-450. It is, however, a mixed-function monooxygenase as demonstrated by $^{18}O_2$ experiments[169]. It is induced by 1,25-$(OH)_2D_3$ and is absent in vitamin D-deficient tissue[170, 171]. Its substrate specificity is similar to that of the 25-OH-D_3-1α-hydroxylase. It has not been further studied nor has it been successfully solubilized at the present time.

3 25-Hydroxyvitamin D_3-26-Hydroxylase

This system is found in kidney mitochondria and is absent in vitamin D-deficient tissue[172]. It is regulated by vitamin D and parathyroid hormone[172]. Little is known beyond these facts except an extra-renal site of 26-hydroxylation is known.

4 Regulation of the Renal 1α- and 24-Hydroxylases

The reaction which is controlled physiologically and which makes the vitamin D system an endocrine system is the 25-OH-D_3-1α-hydroxylase. Since this reaction occurs exclusively in the kidney and since its product has a target of action in intestine and bone it would appear that it fits the criterion of a hormone[8]. In true hormonal fashion the biosynthesis of this metabolite is strongly regulated physiologically by the need for calcium[173, 174] and for phosphorus[175]. In 1971 it was clearly demonstrated that the production and appearance of 1,25-$(OH)_2D_3$ in the rat is markedly stimulated by feeding of low calcium diets and suppressed by feeding a high calcium diet[173]. This effect could be translated into a clear relationship between the serum calcium level and the ability of the animals to produce 1,25-$(OH)_2D_3$[174]. As animals become hypocalcemic or illustrate the need for calcium, the accumulation of 1,25-$(OH)_2D_3$ is stimulated. When calcium is not needed, the accumulation of 1,25-$(OH)_2D_3$ is suppressed. In most situations wherein 1,25-$(OH)_2D_3$ synthesis is suppressed, 24,25-$(OH)_2D_3$ synthesis is stimulated. That the changes in accumulation of 1,25-$(OH)_2D_3$ in plasma result from increased synthesis was shown by direct measurement of the hydroxylases by Omdahl et al.[176] a result confirmed three years later by Henry et al.[176]. Removal of the parathyroid glands eliminates the ability of hypocalcemia to stimulate 1,25-$(OH)_2D_3$ production and stimulates production of 24,25-$(OH)_2D_3$[178]. The administration of parathyroid hormone will restore the ability to make 1,25-$(OH)_2D_3$ and suppress the production of 24,25-$(OH)_2D_3$[178]. Experiments in chicks carried out by Fraser and Kodicek have confirmed that the parathyroid hormone and the parathyroid glands are the primary controlling factor in determining synthesis of 1,25-$(OH)_2D_3$ in response to the need for calcium[179]. Parathyroid control of 1α-hydroxylase in the vitamin D-deficient chick has also been elegantly demonstrated[180] and evidence that this mechanism involves cyclic AMP has been offered[181]. Certainly the control of the renal 1- and 24-hydroxylases by serum calcium is primarily mediated by the parathyroid glands.

Attempts by several investigators to show direct effects of parathyroid hormone on isolated chick renal tubules have been largely unsuccessful[182–184]. This is probably related to the fact that *in vivo* the changeover of the hydroxylases in response to parathyroid hormone or a change in serum calcium is a slow process involving many hours[178, 179]. Undoubtedly isolated renal tubules do not survive for the periods of time necessary to permit expression of the *in vivo* stimulation of 1α-hydroxylase by parathyroid hormone. There has been a report that excessive amounts of parathyroid hormone actually suppress 1,25-$(OH)_2D_3$ production, whereas large amounts of calcitonin, another calcium related hormone, will increase 1,25-$(OH)_2D_3$ production[185, 186]. However, these experiments were carried out in intact animals which could have brought about secondary responses of the parathyroid gland or "c" cells of the thyroid, complicating interpretation of the results. Lorenc et al. demonstrated that this is certainly the case for the calcitonin experiments[187]. Thus the physiologic meaning of these experiments remains in doubt and no evidence is available to support in the idea that calcitonin regulates the vitamin D hydroxylases.

In an attempt to understand the mechanism whereby calcium and the parathyroid complex controls the 1-hydroxylase system, studies involving the addition of calcium and/or parathyroid hormone and other ions to isolated mitochondria, isolated renal tubules, and other *in vitro* systems have been carried out[188–191]. Parathyroid hormone added to isolated renal mitochondria has no effect on the 1-hydroxylase system. Calcium can either stimulate or suppress the 1-hydroxylase system in isolated mitochondria depending upon the nature of the isolation medium for the mitochondria. When they are isolated from EDTA media, calcium will stimulate[191] and when isolated from normal media, the calcium will inhibit[184, 190]. When the components of the 1-hydroxylase system are solubilized and isolated as described above and recombined, the reconstructed system is insensitive to calcium additions[164]. Furthermore, swollen mitochondria carry out the 1-hydroxylation even in the presence of 10 mM calcium[162]. Thus the 1-hydroxylase system itself is not inhibited by calcium ions. It seems likely that the inhibition by calcium *in vitro* is related to altered mitochondrial structure and the inhibition of NADPH production. Whether or not this represents a true physiological control mechanism is in doubt. Certainly many hours are required for the hydroxylases to change over in response to the parathyroid hormone or in response to changes in serum calcium concentration. Were this the result of ionic inhibition or activation, one would expect that these hydroxylases would change very rapidly. Thus the physiologic meaning of the inhibition of 1-hydroxylation in intact mitochondria by calcium is unclear. Experiments which show that the hydroxylases of the kidney have half lives of the order of 2 hr[192, 193] suggests that the ionic environment of the cell may well regulate the biosynthesis or degradation of the enzymes or a controlling protein and that this is the likely mechanism rather than a direct ionic inhibition or activation of the hydroxylases. That this is the case is supported by recent experiments using monkey kidney cell cultures in which 25-OH-D_3-24-hydroxylase is activated by high calcium in the medium and suppressed by parathyroid hormone *in vitro*[194].

Vitamin D administration to vitamin D-deficient rats[170] and chicks[171] suppresses the 1-hydroxylase and stimulates the 24-hydroxylase. This control is exerted by 1,25-$(OH)_2D_3$ itself and the administration of 1,25-$(OH)_2D_3$ to vitamin D-deficient rats brings about a suppression of the 1-hydroxylase and a concommitant stimulation of the 24-hydroxylase[171, 195]. In fact, current work suggests that the 1,25-$(OH)_2D_3$ induces the formation or appearance of the 24-hydroxylase system[171]. These changes apparently involve RNA and protein synthesis[196, 197] and 1,25-$(OH)_2D_3$ does stimulate RNA synthesis in kidney tissue[198]. Nevertheless, the mechanism and nature whereby the renal hydroxylases are regulated by 1,25-$(OH)_2D_3$ remain unknown.

Besides calcium and 1,25-$(OH)_2D_3$, phosphorus plays an important role in the regulation of 1,25-$(OH)_2D_3$ levels in blood. Maintaining rats on a low phosphorus diet results in a marked accumulation of 1,25-$(OH)_2D_3$ in blood even in the absence of the parathyroid gland[175]. Furthermore, phosphate deprivation in thyroparathyroidectomized rats will bring about a stimulation of 1,25-$(OH)_2D_3$ accumulation in blood and tissue. Thus in thyroparathyroidectomized rats a clear inverse relationship between serum inorganic phosphorus levels and 1,25-$(OH)_2D_3$ accumulation has been demonstrated[175]. Under conditions of low blood phosphorus, 1,25-$(OH)_2D_3$

accumulation is stimulated, whereas under conditions of normal or high blood phosphorus, 24,25-$(OH)_2D_3$ is accumulated. Because parathyroid hormone blocks renal reabsorption of phosphate, the idea that this might be the mechanism whereby parathyroid hormone stimulates 1,25-$(OH)_2D_3$ production appeared[175]. A measurement of renal cortical levels of inorganic phosphorus showed a correlation between that level and the ability to produce 1,25-$(OH)_2D_3$. Thus phosphate deprivation or parathyroid hormone administration causes a reduction in renal cortical levels of inorganic phosphorus, whereas conditions which favor 24,25-$(OH)_2D_3$ synthesis increase renal cortical levels of inorganic phosphorus.

Recently a report has appeared which suggested that the renal 25-OH-D_3-1α-hydroxylase is not stimulated by phosphate deprivation[177]. However, an examination of those results revealed that the diet used was not low in phosphorus nor was the length of time to which the animals were exposed to the low phosphorus diet sufficient to stimulate the 1-hydroxylase. Low phosphorus diets do stimulate chick renal 25-OH-D_3-1α-hydroxylase activity about 5-fold[199], although the stimulation is not comparable to that produced by low calcium diets[199]. In addition, it seems quite likely that phosphate deprivation also stimulates the intestine directly inasmuch as rats maintained on 1,25-$(OH)_2D_3$ as their sole source of vitamin D still show a marked elevation of intestinal calcium transport under conditions of phosphate deprivation[200]. A similar conclusion was reached using an analog of 1,25-$(OH)_2D_3$, namely dihydrotachysterol$_3$[209].

It is clear that the biochemical mechanisms whereby the vitamin D hydroxylases of the kidney are regulated have not been solved. It is nevertheless also clear that the need for calcium or the need for phosphorus will stimulate the production of 1,25-$(OH)_2D_3$, a hormone which functions both in calcium mobilization and in phosphate mobilization. Additionally, calcium regulation is mediated for the most part by the parathyroid hormone. Finally, 1,25-$(OH)_2D_3$ itself plays an important role in this regulation by inducing the 24-hydroxylase and suppressing the 1-hydroxylase.

It may seem disturbing that 1,25-$(OH)_2D_3$ can be considered a hormone with two signals (low calcium and low phosphorus) and two functions[8] (calcium mobilization and phosphate mobilization). It would appear that a specific correction of the signal would not be possible. However, the calcium signal causes parathyroid secretion; thus parathyroid hormone accompanies 1,25-$(OH)_2D_3$ in this circumstance, permitting mobilization of bone calcium and renal reabsorption of calcium. The effect of 1,25-$(OH)_2D_3$ on serum phosphate is negated by the parathyroid hormone induced loss of phosphate in urine. The composite effect of the low calcium signal is to elevate serum calcium but not phosphate[8].

The low serum phosphate signal causes 1,25-$(OH)_2D_3$ to appear without parathyroid hormone. Bone calcium is not readily mobilized without parathyroid hormone and calcium is not completely reabsorbed in the kidney. However, phosphate which is absorbed in intestine in response to 1,25-$(OH)_2D_3$ is totally reabsorbed in the kidney[202]. The net effect is to increase serum phosphorus but not calcium[8]. Thus 1,25-$(OH)_2D_3$ can be a dual purpose hormone with two different signals.

5 The Calcium Homeostatic Mechanism (Fig. 5)

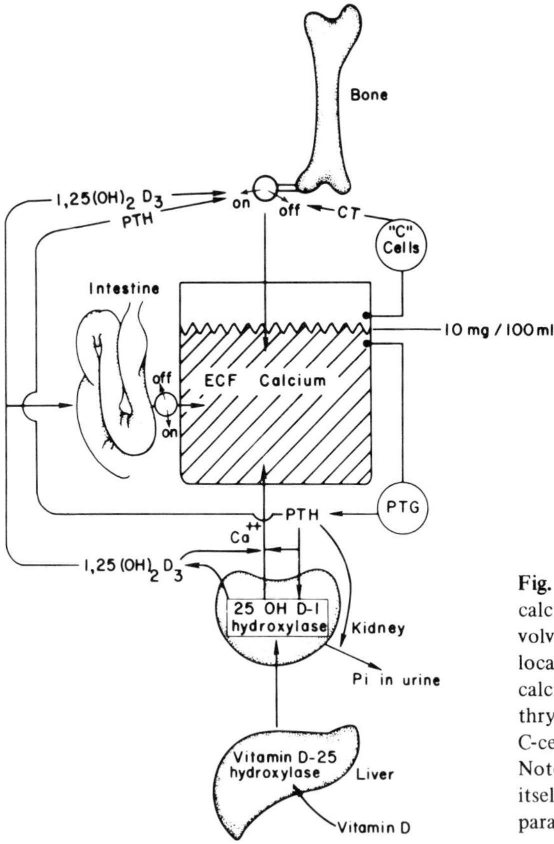

Fig. 5. Diagramatic representation of calcium homeostatic mechanisms involving the vitamin D endocrine system located in the kidney. Note that the calcium sensing organs are the parathryoid glands (hypocalcemia) and the C-cells of the thyroid (hypercalcemia). Note that the vitamin D hormone acts by itself on the intestine and together with parathyroid hormone in bone and kidney

It is now clear that the vitamin D endocrine system is a major factor in the control of plasma calcium and the overall calcium economy of terrestial vertebrates[203]. The parathyroid glands monitor calcium concentration of the plasma (Fig. 5) and in response to low blood calcium secrete the parathyroid hormone[204]. Parathyroid hormone is taken up by the kidney and bone[205]. In the kidney, parathyroid hormone stimulates production of 1,25-$(OH)_2D_3$[178, 179]. The 1,25-$(OH)_2D_3$ then stimulates intestinal absorption of calcium and together with parathyroid hormone stimulates calcium mobilization from bone[205] and renal reabsorption of calcium. These three sources of calcium restore plasma calcium to normal, shutting off parathyroid hormone secretion which in turn shuts down 1,25-$(OH)_2D_3$ synthesis.

Since 1,25-$(OH)_2D_3$ production responds slowly to parathyroid hormone[178, 179] a more quickly-reacting system is needed to prevent low calcium tetany. Parathyroid hormone secretion and action is very rapid as is the lifetime of the hormone itself[207]. Thus parathyroid hormone can, with endogenous 1,25-$(OH)_2D_3$, stimulate the mobilization of calcium from bone and renal reabsorption of calcium but *not* intestinal absorption[205]. Nevertheless, plasma calcium rises but essentially at the expense of bone. This prevents hypocalcemic tetany but uses bone to do so. This could then result in an osteopenia which may precipitate the disease osteoporosis. To protect against this, chronic need for calcium (slight hypocalcemia) causes a prolonged increase in parathyroid hormone levels resulting in increased 1,25-$(OH)_2D_3$ production[178, 179, 200]. This then stimulates intestinal calcium absorption, thereby preventing loss of skeleton and providing for overall economy of calcium in the organism from the environment. This important adjustment is essential to the protection of the skeleton for organisms not living in an environment containing a constant level of calcium.

6 Regulation of the Renal Hydroxylases by Other Factors

It is well known that strontium brings about a vitamin D-resistant rickets and markedly reduces intestinal calcium absorption[208]. It is now clear that the feeding of strontium represses the renal 25-OH-D_3-1α-hydroxylase and that the administration of 1,25-$(OH)_2D_3$ to strontium-fed animals will restore their ability to absorb calcium[209, 210]. Thus at least in part the metabolic basis for strontium-induced rickets has been solved. Recent results suggest that strontium suppresses the 1-hydroxylase by suppressing parathyroid secretion[211].

The administration of large amounts of ethane-1-hydroxy-1,1-diphosphonate (EHDP) has been shown to cause vitamin D-resistant rickets[212] and a markedly repressed intestinal calcium absorption. This has also been shown to be due to a repressed 1,25-$(OH)_2D_3$ biosynthesis[213, 214]. The repressed intestinal absorption brought by the large amount the EHDP can be completely reversed by the exogenous administration of 1,25-$(OH)_2D_3$[215]. It is unknown how the EHDP blocks the 1-hydroxylase, but it is not the result of direct inhibition[216].

Although nephrectomized rats cannot form 1,25-$(OH)_2D_3$, homogenates of rat kidneys cannot carry out significant amounts of the 25-OH-D_3-1-hydroxylation *in vitro*[217]. This contrasts very sharply with the fact that renal homogenates and isolated mitochondria from chicks and Japanese quail clearly carry out the hydroxylation step. The basis for this is the presence of a heat labile inhibitor in rat tissues, which when added to chick kidney preparations will also repress their ability to carry out the 1-hydroxylation[217]. This inhibitor has been found in large amounts in rat blood, intestine and kidney fractions. This inhibitor has been isolated from rat blood and shown to be a protein with a molecular weight of the order of 52,000[218]. This protein is the rat plasma transport protein for 25-OH-D_3[219]. It combines with other proteins in the cytosol of kidney artifactually to form 5–6S binding proteins[220]. By adding a large excess of substrate, this inhibition can be overcome allowing measurement of the renal hydroxylase in mammals (Y. Tanaka and H. F. DeLuca, unpublished results).

Calcium demands are highest during periods of rapid growth, pregnancy and lactation. It is attractive to consider that during these periods increased amounts of 1,25-$(OH)_2D_3$ are made. Blood levels of 1,25-$(OH)_2D_3$ have been reported to be high during these periods[221, 222]. It has also been reported that growth hormone and prolactin stimulate the 25-OH-D_3-1-hydroxylase[223]. These reports remain to be verified and a determination made as to whether these are direct mechanisms or are mediated by the calcium and phosphorus regulating mechanisms.

During egg shell formation in birds, there is great demand for calcium[224, 225]. To provide for these demands, intestinal calcium absorption is increased[225, 226] and medullary bone is resorbed[224, 225]. Medullary bone is the storage system for calcium used by the bird during its reproductive phase of life[226, 227]. It is "woven" bone having little structural capability and forms in response to a direct action by estradiol and testosterone on bone[227]. These two sex hormones are released as the female bird approaches maturity accounting for the initiation of medullary bone formation[227]. Egglaying birds have high levels of the renal 25-OH-D_3-1-hydroxylase activity as compared to males or non-egg-laying mature females[228, 229, 230]. Injection of estradiol to mature male birds stimulates both medullary bone formation and 25-OH-D_3-1α-hydroxylase. Immature birds or castrate males do not respond to estradiol alone, requiring testosterone for response. Progesterone can substitute for testosterone while all three hormones together give marked stimulation of the 1α-hydroxylase[231]. A number of reports have appeared which confirm the stimulation by the sex hormones of 1α-hydroxylase in birds[232, 233] leaving no doubt that modification of the vitamin D endocrine system is involved in the egg-laying and that the sex hormones directly or indirectly affect the renal 25-OH-D_3-1α-hydroxylase. As the female bird approaches maturity, medullary bone is formed but neither the 25-OH-D_3-1α-hydroxylase nor plasma levels of 1,25-$(OH)_2D_3$ are elevated (Fig. 6).

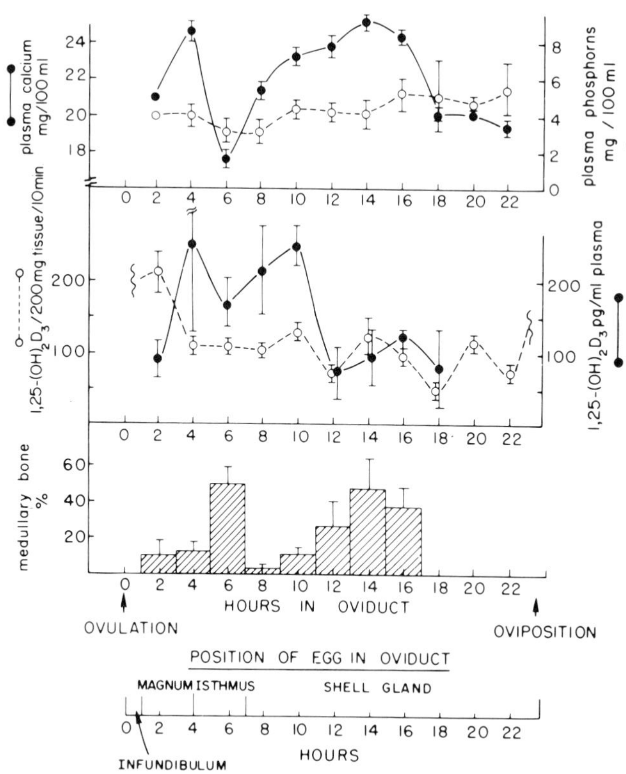

Fig. 6. Sequential changes in the plasma and bone of hens during the egg laying cycle. Note that the plasma 1,25-$(OH)_2D_3$, the renal 25-OH-D-1-hydroxylase, and plasma calcium are always much higher in the egg laying hens than in male or non-egg laying female controls

At the time of ovulation, the 25-OH-D_3-1α-hydroxylase is markedly elevated, and the 24-hydroxylase is suppressed. Plasma 1,25-$(OH)_2D_3$ levels rise to high values and the renal 1-hydroxylase becomes suppressed, while the 24-hydroxylase is elevated. The high levels of 1,25-$(OH)_2D_3$ persist for about 12 hr and then fall to levels still above those in non-egg-laying birds. The formation of medullary bone does not appear related to 1,25-$(OH)_2D_3$ levels rather the 1,25-$(OH)_2D_3$ levels may be involved only to regulate calcium absorption during this period of great calcium need. However, additional work is needed to elucidate the detailed endocrine interactions during the egg-laying cycle.

A possible effect of glucocorticoids on vitamin D metabolism has been published[224)]. Glucocorticoids are known to repress intestinal calcium transport and evidence has been presented which suggests that in fact they do not interfere with production of 25-OH-D_3 or 1,25-$(OH)_2D_3$[235)]. In addition, some work has been presented which suggests that the glucocorticoids repress calcium absorption even in animals given 1,25-$(OH)_2D_3$. More recently Carre et al.[236)] have presented evidence that the glucocorticoids cause the conversion of 1,25-$(OH)_2D_3$ to a more polar and inactive metabolite in the intestine. Unfortunately, this report could not be confirmed. However, low plasma levels of 1,25-$(OH)_2D_3$ have been found in children treated with prednisolone (R. W. Chesney and H. F. DeLuca, unpublished results). Thus some effect of the glucocorticoids on vitamin D metabolism can be anticipated.

7 The Assay of Vitamin D and Its Metabolites

Because of the potential medical applications of the vitamin D metabolites to treatment and in diagnosis of bone disease, considerable interest in measurement of the microquantities of vitamin D and its metabolites in blood has been generated. Vitamin D_2 and vitamin D_3 can be measured by high-pressure liquid chromatography (HPLC) after suitable prepurification and correction for losses during prepurification (R. Shepard, and H. F. DeLuca, unpublished results)[143)]. The vitamin D is detected by ultraviolet absorption using a sensitive monitor at 254 nm. This depends on the cis-triene chromophore having a molar extinction coefficient of 18,200[237)]. 25-OH-D_3 and 25-OH-D_2 can also be measured by HPLC after prepurification[238)]. 24,25-$(OH)_2D$[239)], 25,26-$(OH)_2D$[240)], as well as 25-OH-D[242)] can also be measured by competitive transport protein binding assay but great care must be exercised in purification of each metabolite prior to assay. HPLC purification is required for reliable assays of 24,25-$(OH)_2D_3$ and 25,26-$(OH)_2D_3$[243)]. 1,25-$(OH)_2D$ is measured by radio-receptor competitive binding assay after extraction and a two-step purification including HPLC[244)]. Brumbaugh and Haussler use a two-step purification procedure and use a chromatin binding step in addition to binding to the receptor for measurement[245)].

Recently a multiple assay procedure for determination of all the metabolites of vitamin D has been described[240)]. A flow sheet of the analysis is shown in Fig. 7. Plasma vitamin D metabolite levels in normal and anephric human subjects are shown in Table, 1.

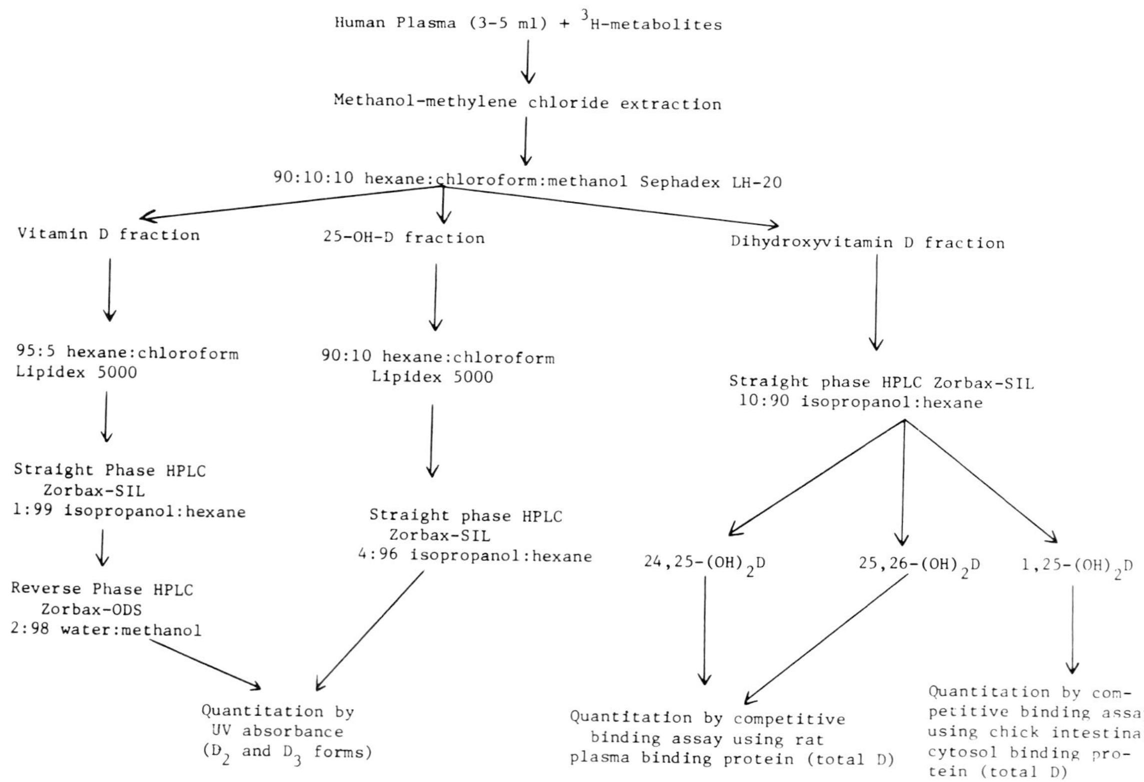

Fig. 7. Flow sheet of analysis of human plasma for vitamin D and its metabolites (Shepard, Horst, Jorgensen, and DeLuca, unpublished results)

Table 1. Vitamin D metabolite levels in normal and anephric man

Metabolite	Plasma Concentration (ng/ml)[a] Normal	Normal with high exposure to sun	Anephric
D_2	1.2 ± 1.4	1.0 ± 0.7	1.3 ± 0.6
D_3	2.3 ± 1.6	26.1 ± 7.2	1.4 ± 0.6
Total D	3.5 ± 2.5	27.1 ± 7.9	2.7 ± 0.8
25-OH-D_2	3.9 ± 3.1	1.3 ± 0.4	18.7 ± 9.0
25-OH-D_3	27.6 ± 9.2	55.5 ± 3.8	17.7 ± 12.2
Total 25-OH-D	31.6 ± 9.3	56.8 ± 4.2	36.4 ± 16.5
24,25-$(OH)_2$D	3.5 ± 1.4	4.3 ± 1.6	1.9 ± 1.3
25,26-$(OH)_2$D	0.7 ± 0.5	0.5 ± 0.2	0.6 ± 0.3
1,25-$(OH)_2$D	0.031 ± 0.009		n.d.[b]

[a] Mean ± S.D. [b] Not detectable.

8 Use of Vitamin D Metabolites in Treatment of Disorders of Bone and Calcium Metabolism

It is obvious that a variety of disorders would result from a disturbance of the vitamin D endocrine system. Fat malabsorption would result in a deficiency of vitamin D giving rise ultimately to osteomalacia or rickets or secondary hyperparathyroidism. A hepatic disorder such as severe cirrhosis, or biliary atresia, may result in malabsorption of vitamin D and defective vitamin D-25-hydroxylation. Dilantin and phenobarbital cause low plasma 25-OH-D levels resulting in rickets and osteomalacia[246]. A lack of parathyroid glands would cause a severe hypocalcemia and tetany. Kidney failure results in severe secondary hyperparathyroidism, osteomalacia and osteosclerosis[247]. These patients cannot produce 1,25-$(OH)_2D_3$ when in need of calcium. Vitamin D dependency rickets is likely a defect in 25-OH-D-1α-hydroxylation[248]. Osteoporosis may in part result from inadequate 1,25-$(OH)_2D_3$ production and thus inadequate calcium absorption resulting in loss of bone[249, 250].

The application of 1,25-$(OH)_2D_3$, its analog 1α-OH-D_3, and 25-OH-D_3 to these diseases is, therefore, evident. Certainly 1,25-$(OH)_2D_3$ or 1α-OH-D_3 would be effective in renal osteodystrophy, hypoparathyroidism, vitamin D dependency rickets, and drug induced osteomalacia. It is very promising as a treatment for various forms of osteoporosis. A review of the clinical literature related to this is beyond the scope of this chapter, but interested readers are directed elsewhere for this information[222, 249–252].

VI Mechanism of Action of 1,25-Dihydroxyvitamin D_3 (Fig. 8)

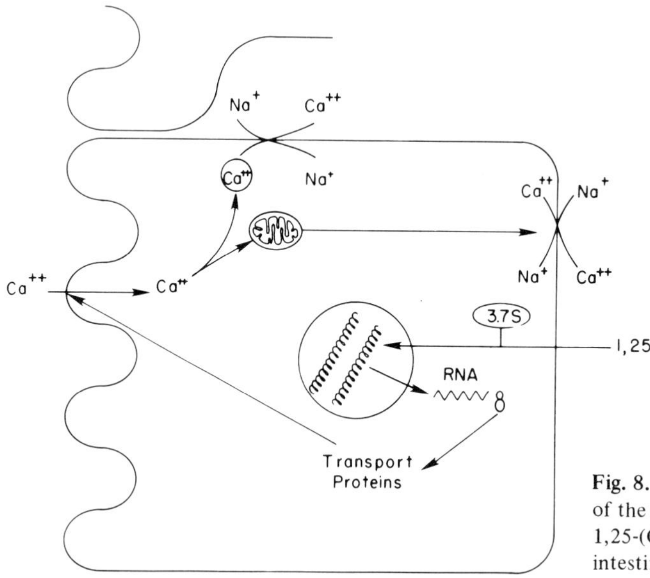

Fig. 8. Diagramatic representation of the mechanism whereby 1,25-$(OH)_2D_3$ may initiate intestinal calcium transport

There are clearly two known sites of action of 1,25-$(OH)_2D_3$: intestine and bone. There is additional evidence that 1,25-$(OH)_2D_3$ is active on kidney and there is indirect evidence that a metabolite of vitamin D may be active on muscle and on parathyroid glands. By far the most thoroughly studied is the role of vitamin D in the intestine although some evidence is available concerning the mechanism of action of 1,25-$(OH)_2D_3$ in bone.

1 Mechanism of Action in the Intestine

It is well known that vitamin D stimulates intestinal calcium absorption. Through the work of several groups has come the understanding that in response to vitamin D, calcium is transported against an electrochemical potential gradient in the small intestine[253–256]. The most rapid rate of calcium transport is in the duodenum followed by jejunum and ileum[257]. Harrison and Harrison have also shown clearly that vitamin D improves intestinal calcium transport even in the colon[258]. Physiologically, because of the time during which calcium is subject to absorption, it seems evident that the distal portions of the small intestine are primarily responsible for the bulk of intestinal calcium absorption.

The movement of calcium against an electrical and concentration gradient requires metabolic energy[254–257]. The site of vitamin D activation of the calcium transport process is not entirely settled, although there is some agreement that vitamin D in some way alters the characteristics of the brush border membrane to permit entry of calcium but the evidence on this point is equivocal[255, 256, 259, 260]. The transfer of calcium across brush borders, whether an active process or not has not been entirely settled, but from a thermodynamic point of view, one would predict that it is not active since intracellular calcium concentration is below that in intestinal lumen[261]. Schachter and his colleagues have provided evidence that vitamin D also improves the transfer of calcium across the serosal surface of the intestinal epithelial cell[260]. However, this result is also not conclusive.

The response of intestinal calcium transport to vitamin D is sensitive to the pretreatment by actinomycin D[262]. However, it is unclear whether actinomycin D blocks the intestinal response to 1,25-$(OH)_2D_3$[263, 264]. There is some evidence which suggests that the lifetime of the 25-OH-D_3-1-hydroxylase and its messenger is sufficiently short that the apparent block in vitamin D action on intestinal calcium transport may be due to the decay of the messenger and of the enzyme for the 1-hydroxylation reaction in the kidney[265]. Once the 1-hydroxylase is bypassed by the administration of 1,25-$(OH)_2D_3$, actinomycin D does not block the transport process in the rat at least[263]. This finding has been confirmed and extended[264]. Tsai et al. have provided conflicting evidence in the chick that actinomycin D does block the intestinal response to 1,25-$(OH)_2D_3$[193]. However, they found it necessary to administer the actinomycin D every 2 hr before a block could be effected. Whether this is a bona fide block or a toxic reaction to the antibiotic remains undetermined. In any case the use of RNA and protein synthesis inhibitors *in vivo* to deduce the mechanism of action of 1,25-$(OH)_2D_3$ results in unclear and difficult to interpret results, especially if RNA and protein synthesis is incompletely blocked[264].

Wasserman and his colleagues have discovered a calcium binding protein which appears in the intestine in response to vitamin D[266]. This calcium binding protein has a molecular weight of 28,000 in the chick and 8,000–12,000 in the rat and other mammalian species[266]. This protein binds two four calcium ions per mole and its appearance and concentration correlate approximately with the rate of intestinal calcium transport under a number of physiological conditions[266]. Inasmuch as this protein binds calcium, appears only after vitamin D administration and correlates approximately with intestinal calcium transport, it has been surmized that it must play a role in the transport process[266–268]. However, there are conditions under which the correlation between calcium binding protein and intestinal transport is quite poor. For example, cortisone inhibits intestinal calcium transport while it stimulates calcium binding protein production[269]. Intestinal calcium transport precedes the appearance of calcium binding protein in response to vitamin D as measured by the Chelex method[270]. The amount of calcium binding protein in the intestine at any time in response to vitamin D does not correlate well with the calcium transport phenomenon[270]. More recent evidence with 1,25$(OH)_2D_3$ and a sensitive immunoassay confirm this result[271, 272]. Further, calcium binding protein production can be blocked by protein synthesis inhibitors while the intestinal absorption process is not[264]. Finally, it can be shown that following 1,25-$(OH)_2D_3$ administration, the intestinal calcium transport response of chicks decays to preinjection level while the calcium binding protein level remains high[273]. It is, therefore, evident that the question of whether calcium binding protein participates in calcium transport has not been entirely settled and to be sure there has not been an experiment which demonstrates clearly the relationship of calcium bindung protein to the calcium transport process. At the very least, 1,25-$(OH)_2D_3$ must influence calcium transport in intestine at least in part by some mechanism other than stimulating calcium binding protein production.

Corradino has developed an intestinal organ culture system using embryonic chick intestine[79, 274]. Before hatching, chick embryonic intestine does not contain calcium binding protein. When this intestine is cultured in the presence of vitamin D or its metabolites, calcium binding protein appears as measured by immunoassay. Furthermore, the ability of intestine to bind calcium increases. In this system even vitamin D is effective although much less effective than 25-OH-D_3, which is in turn less effective than 1,25-$(OH)_2D_3$ in stimulating the appearance of calcium binding protein[79, 274]. In this system, both actinomycin D and α-aminitin, the RNA polymerase II inhibitor, prevent the calcium binding protein response to vitamin D metabolites[275]. There is little doubt that the appearance of calcium binding protein in this system is an induction phenomenon, but whether it is related to the response of young growing rats or chicks to vitamin D metabolites is unresolved. In regard to the induction of calcium binding protein in small intestine, it has been demonstrated that chick embryo possesses the ability to metabolize vitamin D to 25-OH-D_3 and also to 1,25-$(OH)_2D_3$[276], but the intestine prior to hatching does not possess the ability to respond to 1,25-$(OH)_2D_3$ and more recent evidence suggests that these small intestines lack a 3.5S receptor for 1,25-$(OH)_2D_3$. Thus the failure to induce calcium binding protein prior to hatching may be due to a lack of the receptor machinery for 1,25-$(OH)_2D_3$.

A fluorescent antibody study of the calcium binding protein location has suggested that it is formed in the goblet cells and is secreted from the goblet cells along the surface of the columnar epithelial tissues[277]. A wider dispersion of this protein has been suggested from other immunofluorescent antibody studies[278], but the question of specificity of antibody must be taken into account. The exact cellular and subcellular localization of calcium binding protein is as of this date unresolved.

Work with radioactive vitamin D metabolites has shown that the nuclear fraction contains as much as 80% of the administered radioactive 1,25-$(OH)_2D_3$[279–282]. However, it is not at all certain whether all of this radioactivity is contained within the pure nuclei. Unfortunately methods of preparing pure nuclei from other tissues have not been successfully applied to the intestine. Even the method specifically developed for this purpose results in a low yield of intestinal nuclei of the order of 20 to 40%[279]. Recently the synthesis of high specific activity tritiated 25-OH-D_3 has been completed[283], converted to 1,25-$(OH)_2D_3$ and used to study subcellular localization by frozen section autoradiography and physiological doses[284]. Specific nuclear location of 1,25-$(OH)_2[^3H]D_3$ could be demonstrated in intestinal villus and crypt cells but not in muscle, liver, submucosa and most of the kidney (Fig. 8). This localization preceded the initiation of intestinal calcium absorption by 1,25-$(OH)_2D_3$[284]. Thus at least a portion of the mechanism of action of 1,25-$(OH)_2D_3$ must involve nuclear function.

Working under the assumption that the nuclear location of 1,25-$(OH)_2D_3$ is of functional importance, emphasis has been placed on the idea that 1,25-$(OH)_2D_3$ in the intestine functions in a manner similar to other steroid hormones. It has been suggested that 1,25-$(OH)_2D_3$ becomes bound to a cytosolic receptor protein which undergoes change as it enters the nucleus and receptor bound steroid activates the transcription of specific genomes which code for functional calcium and phosphorus transport proteins. In mammalian species, Hadded et al. have studied cytosolic proteins which bind $1,25\text{-}(OH)_2D_3$[285–287]. They report the widespread existence of a 6S protein which binds both $25\text{-}OH\text{-}D_3$ and $1,25\text{-}(OH)_2D_3$. It was suggested that this protein is a receptor for $25\text{-}OH\text{-}D_3$ and that $25\text{-}OH\text{-}D_3$ has widespread function. The existence of this protein in both target and nontarget tissues was confirmed[288] but shown to be an artifactual combination of serum transport protein and a cytosolic protein[220, 289]. It is, therefore, not a receptor as previously suggested. A $1,25\text{-}(OH)_2D_3$ receptor has been reported for the parathyroid glands[290] but whether it has a role in this tissue remains unknown since reports on the "direct" effects of $1,25\text{-}(OH)_2D_3$ on parathyroid secretion or peripheral parathyroid hormone have been variable[107, 291, 292]. Certainly calcium can control parathyroid hormone secretion in a vitamin D deficient animal[292].

Haussler and his colleagues have studied a cytosolic and nuclear receptor protein from chick intestine[293–295]. They have reported the existence of a 3.0 to 3.7S protein which binds specifically $1,25\text{-}(OH)_2D_3$. Although difficult to reproduce initially, the existence of this protein in chick intestine was confirmed[288]. This protein has the characteristics which one would consider as indicative of a receptor protein, namely it has high affinity and low capacity for the $1,25\text{-}(OH)_2D_3$. It is not found in nontarget tissues and, most important, under conditions of incubation with nuclei or chromatin, radioactive $1,25\text{-}(OH)_2D_3$ is transferred in a rather specific manner to that fraction[296, 297]. This is believed indicative that $1,25\text{-}(OH)_2D_3$ functions by inducing the transcription of genetic information which in turn codes for the calcium transport system. A 3.2S $1,25\text{-}(OH)_2D_3$ receptor has been shown in rat intestine[298, 299] and rat and chick bone[300]. The chick intestinal receptor is highly specific for $1,25\text{-}(OH)_2D_3$[81, 82]. A purification of it has been reported but is not convincing[301].

Brumbaugh et al. have also shown that their "chromatin binding" method can be used as an assay for $1,25\text{-}(OH)_2D_3$ and have successfully applied it in the study of blood levels of this important metabolite in a variety of diseases[244]. A simplified assay for $1,25\text{-}(OH)_2D_3$ using only the cytosolic $1,25\text{-}(OH)_2D_3$ receptor from chick intestine has also been developed and used as discussed previously[245, 302].

It seems clear that $1,25\text{-}(OH)_2D_3$ must be involved in some induction process in intestine but whether this is an induction of a calcium transport substance or not remains to be established. Lawson and his colleagues have studied the formation of calcium binding protein in cell free preparations from vitamin D-deficient chick intestine and from chicks given vitamin D[303, 304]. The polysomes from chicks given vitamin D possess the ability to form a substance which is immunologically similar to calcium binding protein as compared to preparations from vitamin D-deficient controls. Attempts have been made to show the existence of a specific messenger for the calcium binding protein that is formed in response to the $1,25\text{-}(OH)_2D_3$, but so far evidence for this is far from complete[304]. There seems little doubt that calcium binding protein is assembled in response to $1,25\text{-}(OH)_2D_3$ either directly or indirectly and that the readout of messenger RNA must be involved. However, this does not settle how $1,25\text{-}(OH)_2D_3$ functions in initiating this phenomenon. It also carries with it no guarantee that this will explain the mechanism of action of 1,25-$(OH)_2D_3$ in the calcium transport process. This nevertheless represents an extremely fertile area of investigation which will undoubtedly yield important new results in the near future. Other mechanism have been suggested[305)] but so far have not received substantial support.

It is unknown how calcium traverses the columnar epithelial cell during the transfer process. Likely it is bound to the mitochondria which act as a buffer controlling the ionized calcium level of the cell[306)]. The calcium is likely released at the basement membrane surface and it has been found that the release of calcium at the basement membrane surface requires the presence of sodium ions[259)]. It has been suggested that a sodium gradient in exchange for calcium brings about the final expulsion of calcium from the columnar epithelial cells depleting that portion of the cell of calcium and causing the mitochondria to release its calcium. (Fig. 8).

Besides the role of 1,25-$(OH)_2D_3$ in the transfer of calcium across the intestinal membrane it is known that this substance also activates the transfer of inorganic phosphate across intestinal illeum and jejunum[46, 73, 307–310)]. This process has been shown to be independent of calcium transport and thus represents an entirely different function of the vitamin. Evidence has been presented that 1,25-$(OH)_2D_3$ rather than 25-OH-D_3 functions in this process[73, 310)]. Little is known concerning the mechanism of phosphate transfer except that it is a sodium dependent[307)] and active process[310)]. So far a specific phosphate binding protein has not yet been found.

2 The Role of 1,25-$(OH)_2D_3$ in Bone

It is well known that vitamin D administration results in the alleviation of rickets and osteomalacia, or in biochemical terms stimulates the mineralization of organic matrix of bone and growth plate cartilage. The major defect in the mineralization of bone in vitamin D deficiency appears to be a lack of supply of calcium and phosphorus to the mineralization centers[36, 37, 311)], although a direct effect of some form of vitamin D on mineralization itself cannot be excluded and is in fact quite likely.

Vitamin D functions in the process of calcium mobilization from previously formed bone making it available to the extracellular fluid upon demand by the calcium homeostatic system as described under the regulation of vitamin D metabolism. From the experiments described in the metabolism section it is clear that 1,25-$(OH)_2D_3$, rather than 25-OH-D_3 or vitamin D_3 functions in the mobilization of calcium from bone under physiologic conditions. The mechanism whereby 1,25-$(OH)_2D_3$ initiates mobilization of calcium from bone is not at all understood. In contrast to the response of intestinal calcium transport to 1,25-$(OH)_2D_3$ this process is blocked by the prior administration of actinomycin D suggesting that in fact a transcriptive event is involved in this activation[312)]. As previously pointed out, bone possesses a specific receptor for 1,25-$(OH)_2D_3$[301)]. The nature of the protein or proteins which are formed in response to 1,25-$(OH)_2D_3$ in the bone cells has not yet been determined however. *In vivo* the 1,25-$(OH)_2D_3$ activation of bone calcium mobilization requires the presence of parathyroid hormone[206)] but the nature of this interdependence is not known. Besides mineralization and mobilization of calcium from bone fluid from a calcium homeostatic point of view, two other processes are known in which vitamin D must participate. These are bone modeling and remodeling[313]. Modeling is the process involved in bone growth and shaping whereas remodeling involves resorbstion and replacement with new bone primarily from a repair and maintenance point of view. In this process, bone must be resorbed by osteoclasts and then reformed by osteoblastic activity. Without vitamin D these processes are retarded. The bone resorption aspect of these processes have been successfully studied in culture[76, 77, 314–316]. It is stimulated markedly by 1,25-$(OH)_2D_3$ and parathyroid hormone independently of each other[314, 315]. The nature of this stimulation remains unknown. The formation portion of the process is dependent on a form of vitamin D likely, 1,25-$(OH)_2D_3$ although both 24,25-$(OH)_2D_3$ and 25-OH-D_3 have been suggested to be the active form in this process. Much work ist needed to clarify this important area of vitamin D function.

3 The Function of 1,25-$(OH)_2D_3$ in Other Tissues

There has been evidence presented which suggests that vitamin D increases renal tubular reabsorption of calcium. In early work by Gran it appeared that vitamin D increased the retention of calcium by kidney[317]. However, these experiments did not have the renal physiological functions controlled in a rigorous manner. More recently in experiments designed for the study of influence of vitamin D metabolites on the reabsorption of phosphate in the renal tubules, it has become evident that 1,25-$(OH)_2D_3$ improves renal reabsorption of calcium[318]. More work is needed in this area, however. It appears that 1,25-$(OH)_2D_3$ does have an influence on renal function. In agreement with this, a calcium binding protein in the kidney which is depressed in vitamin D deficiency and which is increased upon adminstration of vitamin D has been demonstrated[319, 320] and some evidence of increase of calcium reabsorption in response to 1,25-$(OH)_2D_3$ has been obtained[321]. Again the mechanism of 1,25-$(OH)_2D_3$ function in the kidney is not known.

Some attention has been focused on a possible role of vitamin D in renal reabsorption of phosphate. Although early work suggested a vitamin D stimulated increase in renal reabsorption of phosphate[322], this appeared the result of an adjustment of parathyroid glands and not a direct effect of vitamin D on renal reabsorption of phosphate. More recent attempts utilizing the metabolites of vitamin D are also not convincing since vitamin D-deficient animals were not used[323, 324], and thus pharmacological effects of the metabolites were studied. Other reports with rats have appeared which suggest but do not prove an effect of vitamin D on the reabsorption of phosphate[325].

It is well known that vitamin D increases serum phosphate levels of rachitic rats which is in turn responsible for the mineralization of bone[326]. The source of the phosphate has been examined in rats on extremely low phosphorus diets so that intestinal absorption would be a minor factor[326]. The response to 1,25-$(OH)_2D_3$ was found to the same extent in these animals as in those on a 0.1% phosphorus diet. Increased renal reabsorption of phosphate was excluded by direct examination[318].

The animals avidly reabsorb all the phosphorus presented to the kidney even without vitamin D. The source of this phosphate proved to be bone as revealed by ^{45}Ca and ^{32}P experiments[327]. Parathyroidectomy did not prevent the response[327]. Thus in hypophosphatemic animals, 1,25-$(OH)_2D_3$ can mobilize calcium and phosphorus from bone without parathyroid hormone.

Table 2. Synthetic metabolites and analogs

Vitamin D analogs	
3-epi D_3 (22a)[360, 361]	(225)-22-OH-D_4 (1f)[386]
5,6-*trans* D_3 (5a)[84]	3-NH_2-D_3 (3a)[374]
5,6-*trans*-3-epi D_3 (6a)[360, 361]	3-epi NH_2D_3 (4a)[374]
2β-OH-D_3[381]	25-aza-D_3 (1s)[346]
4α-OH-D_3[382]	Δ^{24}-D_3 (1u)[346]
20-OH-PC (1r)[383]	Δ^{25}-D_3 (1v)[346]
(24*R*) & (24*S*)-24-OH-D_3 (1b & 1c)[111, 345]	1-F-D_3 (15a)[372]
(10*R*) & (10*S*)-19-OH-DHD_3 (7a & 8a)[369, 370]	25-F-D_3 (1t)[346]
DHT_3 (10a)[368]	cyclo D_3 (11a)[380]
DHT_2 (10d)[368, 369]	4,4-$(CH_3)_2$-D_3[394]
D_4 (1e)[384, 385]	
25-OH-D analogs	
25-OH-D_3 (1g)[49, 95, 328, 329]	24-Homo-25-OH-D_3 (1m)[388]
5,6-*trans*-25-OH-D_3 (5g)[84]	24-Nor-25-OH-D_3-Pentanor-25-OH-D_3 (1n–q)[388]
24,25-$(OH)_2D_3$[330, 333]	27-Nor-25-OH-D_3 (1k)[387, 384]
(24*R*) & (24*S*)-24,25-$(OH)_2D_3$ (1h & 1i)[94, 95]	26,27-Bisnor-25-OH-D_3 (1l)[387, 383]
25,26-$(OH)_2D_3$[117, 118, 331, 344]	5,6-trans-24-nor-25-OH-D_3 (5n)[389]
2,25-$(OH)_2D_3$[156, 387]	
1α-OH-D analogs	
1α-OH-D_3 (13a)[348, 356, 367, 377, 390, 391]	1β-OH-D_3 (16a)[56, 375]
1α,25-$(OH)_2D_3$ (13g)[55, 96, 347, 349, 377]	(24*R*) & (24*S*)-1α,24-$(OH)_2D_3$ (13b & 13c)[345, 352]
1α-OH-D_2 (13d)[351, 377]	(24*R*) & (24*S*)-1α,24,25-$(OH)_3D_3$ (13h & 13i)[345, 350]
1α-OH-PC (13w)[353]	24-Nor-1α,25-$(OH)_2D_3$ (13n)[393, 389]
5,6-*trans*-1α-OH-D_3[377]	25-F-1α-OH-D_3 (13t)[371]
3-deoxy-1α-OH-D_3 (17a)[362, 363]	3-F-1α-OH-D_3 (18a)[373]
3-deoxy-1α,25-$(OH)_2D_3$ (17g)[392]	1α-OH-Cyclo-D_3 (12a)[377]
3-OCH_3-1α-OH-D_3[393]	4,4-$(CH_3)_2$-1α-OH-D_3[394]
1α-OH-3-epi D_3 (16a)[354, 355]	4,4-$(CH_3)_2$-1α-OH-epi-D_3[394]

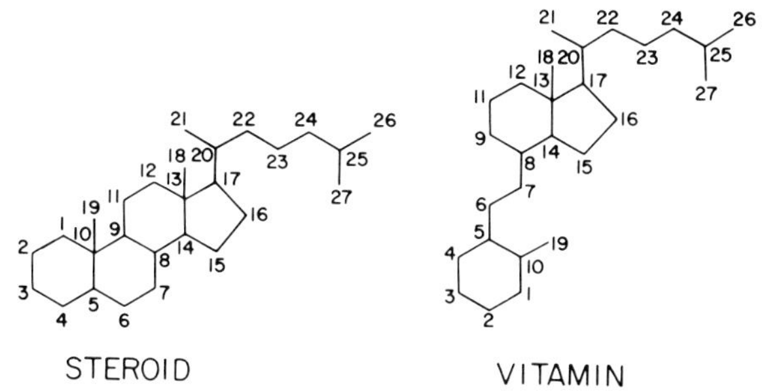

Fig. 9. Steroid and vitamin skeletal numbering

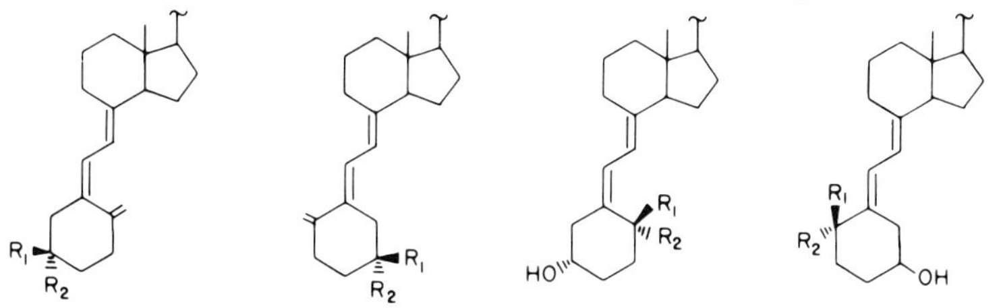

1) R_1 = H, R_2 = OH
2) R_1 = OH, R_2 = H
3) R_1 = H, R_2 = NH_2
4) R_1 = NH_2, R_2 = H

5) R_1 = OH, R_2 = H
6) R_1 = H, R_2 = OH

7) R_1 = CH_2OH, R_2 = H
8) R_1 = H, R_2 = CH_2OH

9) R_1 = CH_3, R_2 = H
10) R_1 = H, R_2 = CH_3

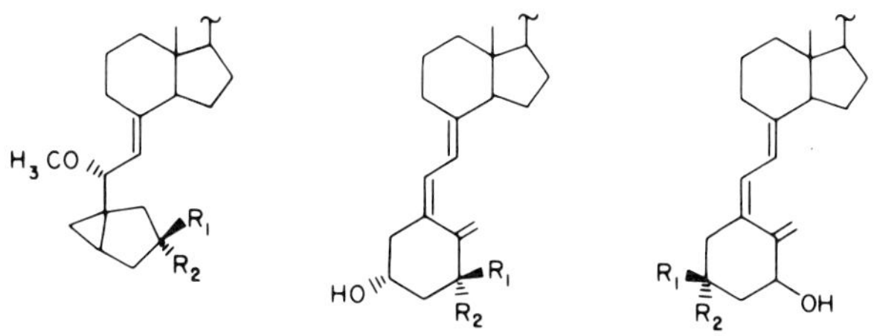

11) R_1 = H, R_2 = H
12) R_1 = OH, R_2 = H

13) R_1 = OH, R_2 = H
14) R_1 = H, R_2 = OH
15) R_1 = F, R_2 = H

16) R_1 = OH, R_2 = H
17) R_1 = H, R_2 = H
18) R_1 = H, R_2 = F

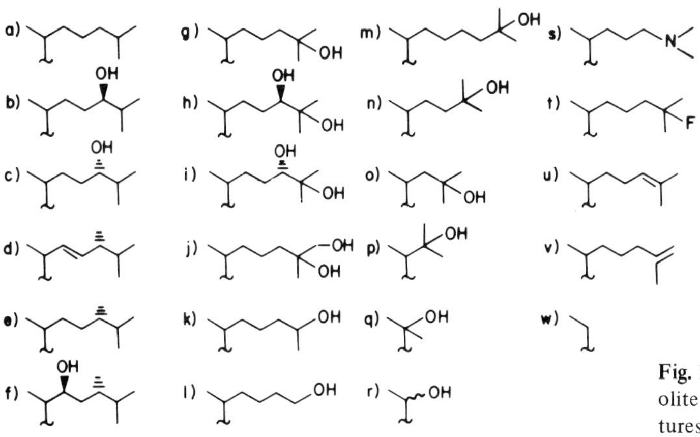

Fig. 10. Synthetic metabolite and analog structures

4 Summary of the Mechanism of Action of Vitamin D and Its Metabolites

Much work has been expended in this area but so far a clear mechanism whereby any of the well known forms of vitamin D function remains undetermined. Most work has been carried out in intestine in which the major subcellular localization of 1,25-$(OH)_2D_3$ is found in the nuclear-debris fraction. A calcium binding protein appears following the administration of vitamin D and its metabolites which has been suggested to play a role in the calcium transport process. The calcium transport process which makes its appearance following vitamin D administration is an active one in which calcium is transported against an electrical and concentration gradient. The primary site of function of vitamin D is likely at the brush border membrane and the evidence for its functioning elsewhere is equivocal. In addition to calcium transport, 1,25-$(OH)_2D_3$ stimulates intestinal phosphate transport by an undetermined mechanism. It also stimulates mobilization of calcium from bone, and increases renal reabsorption of calcium. In addition some form of vitamin D increases muscle strength by an as yet undetermined process. It is in the area of mechanism of action that one might expect marked investigational activity which hopefully will bring about the elucidation of the cellular and molecular mechanisms of this vitamin's (hormone's) action.

VII Synthesis of Vitamin D Compounds

1 Introduction

The elucidation of the biochemistry of the vitamin D endocrine system has led to a renewed interest in the synthetic aspects of the field. Because of the limited availability of vitamin D metabolites from natural sources, the production of large quantities of metabolites and analogs is essential for detailed biomedical experimentation. Vitamin D analog synthesis has not only aided in determining the structural and stereochemical parameters of the natural metabolites, but has also helped to establish a foundation for structure-activity theories.

The energy expended in this area is reflected by the large number of structurally diverse analogs compiled in Table 2. Basic synthetic approaches to metabolites and analogs of vitamin D consist of distinct side chain and A-ring modification sequences which have classically been performed on a steroidal precursor. The suitably functionalized derivative is then converted to the 9,10-secosteroid, vitamin D, by photochemical and thermal isomerization of the corresponding provitamin (i. e. 5,7-diene). Some recent synthetic endeavors offer an attractive alternative to this scheme by attempting to modify directly a pre-formed vitamin nucleus to the analog of choice.

2 Side Chain Elaboration

Construction of side chain derivatives has usually begun from readily available molecules such as aldosterone, pregnenelone, and the C-22 aldehyde produced from the ozonolysis of sigmasterol or diene-protected ergosterol. Other more convenient starting materials have been employed (i. e. 25-carbomethoxy, 25-keto, and 24,25-ene-type side chains), but being difficult and expensive to obtain these key intermediates represent synthetic targets in their own right. Schemes (1)–(3) summarize

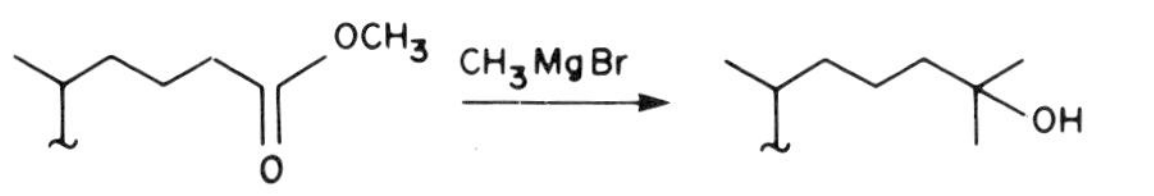

(1)

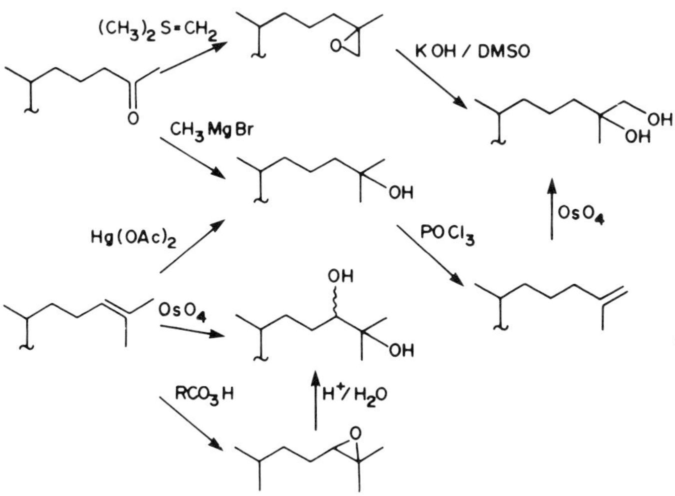

(2)

(3)

typical conversions yielding 25-hydroxy-, 24,25-dihydroxy-, and 25,26-dihydroxy cholesterol derivatives from homocholenic acid ester [scheme (1)][55, 328], 27-nor-25-ketocholesterol [scheme (2)][49, 93, 117, 329–331], or desmosterol [scheme (3)][94, 332–334]. Direct photooxygenation of the cholesterol side chain provides another reasonably efficient route to 25-hydroxy derivatives[335].

An effective 10-step procedure leading to a 25-OH side chain in 42% yield from the C_{19} steroid aldosterone has been accomplished [scheme (4)][336]. The addition of a two-carbon fragment via a Reformatsky reaction, dehydration, and selective reduction are the initial reactions. Further elaboration proceeds by a stereospecific alkylation with the ethylene ketal of 5-bromo-2-pentanone in the presence of lithium diisopropyl amide followed by conversion of the 21-ester to the corresponding 21-methyl compound. Acid catalyzed removal of the ethylene ketal group affords the critical 27-nor-25-keto intermediate.

From the C_{21} steroid, pregnenolone, many excellent synthetic schemes to hydroxylated side chain derivatives have been developed. Direct condensation of the

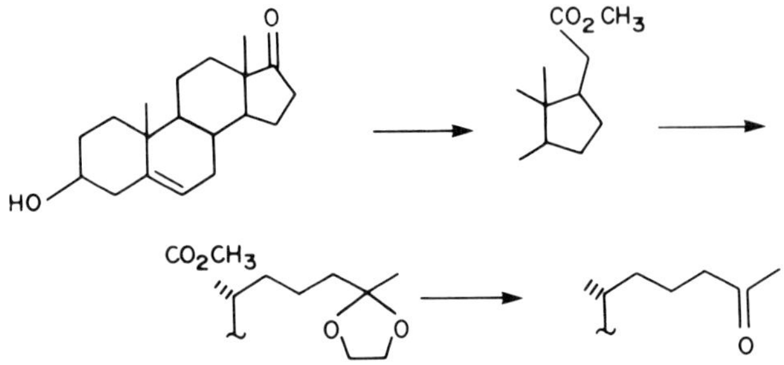

(4)

C-21 keto group with the Wittig reagent derived from 5-bromo-2-pentanone ethylene ketal [scheme (5)][337] gives the expexted 20,22-olefin in good yield. Removal of the ketal protecting group and selective hydrogenation of the side chain double bond results in a 9:1 mixture of C-21 *R* and *S* isomers (62% from pregenolone) which are separated by fractional crystallization.

Another creative scheme which originates from pregnenolone [scheme (6)][338] utilizes vinyl magnesium bromide as the initial reactant and yields the corresponding

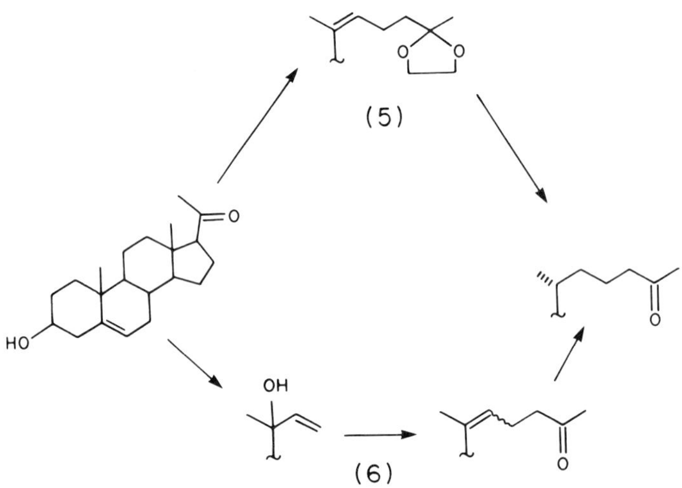

allylic alcohol. Collidine catalyzed reaction with diketene in refluxing decalin produced the *cis*- and *trans*-keto olefins by an interesting pathway consisting of internal nucleophilic attack of an intermediate enolate ion followed by (or in concert with) decarboxylation of the resultant β-keto acid. Catalytic reduction and fractional crystallization again gives the (20*S*)-25-ketone in 30% yield.

By far the most widely used starting material for side chain elaboration schemes has been the C-22 aldehyde. This compound, readily obtained from the ozonolysis of sigmasterol i-ether or the 4-phenyl-1,2,4-triazoline-3,5-dione (PTAD)-diene adduct of ergosterol[339, 340], has been modified in a number of ways.

Reduction to the alcohol, tosylation and conversion to the C-22 iodo derivative provides the substrate for reaction with α-(dimethyl allyl) nickel bromide. This coupling reaction gives the desmosterol side chain directly [scheme (7)][341].

Desmosterol-type side chains have also been constructed from C-22 halide precursors by modified Grignard reactions[342]; however, yields are usually low.

Treatment of the 22-iodo compound with the appropriate lithium acetylide [scheme (8)][339] and catalytic reduction represents another efficient route to the 25-hydroxylated side chain.

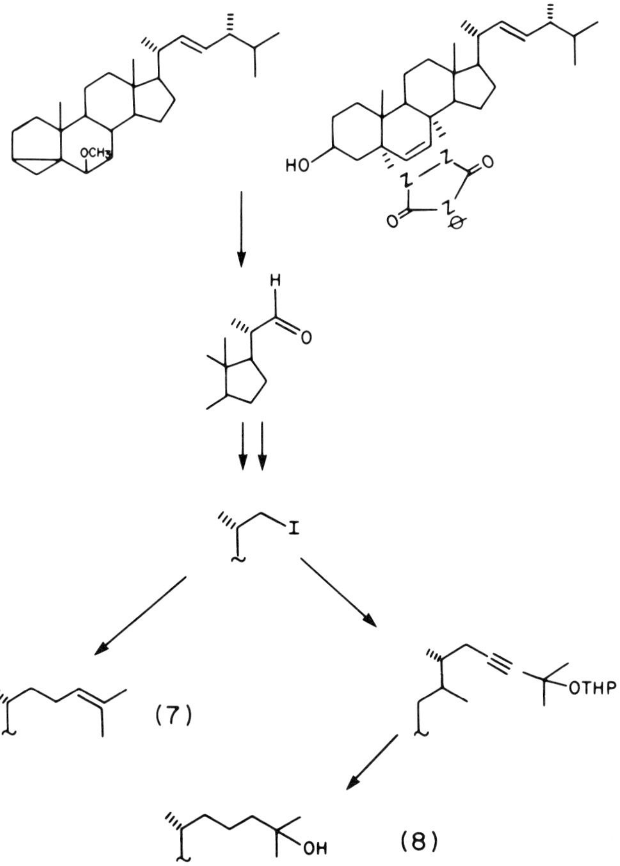

(7)

(8)

The PTAD-diene protected 22-aldehyde analog has been used by Williams and Eyley to synthesize 24,25-dihydroxy derivatives via an aldol reaction with the tetrahydropyranyl ether of 3-hydroxy-3-methyl-2-butanone followed by $NaBH_4$/pyridine reduction [scheme (9)][340]. The same starting aldehyde is also a precursor for synthesis of the 25,26-dihydroxy-5,7-diene [scheme (10)][340].

Most of the reported routes to the 24,25- and the 25,26-dihydroxy analogs yield epimeric mixtures, two exceptions being the Hoffmann-La Roche synthesis of the 24,25-diols[343] and Ikekawa's synthesis of the 25,26-diols[344]. The 24,25-

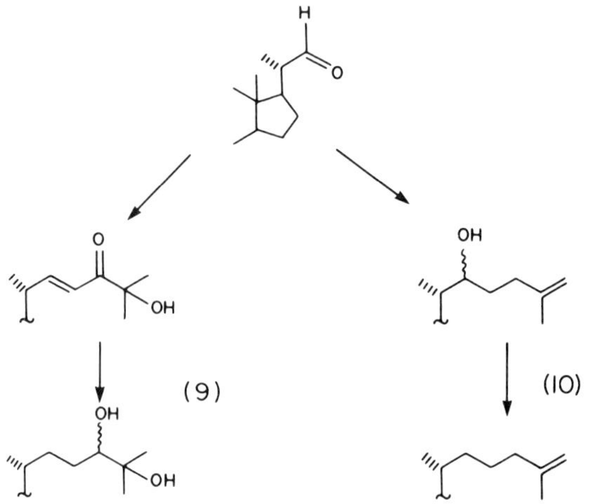

epimeric diols have been chromatographically separated[332] and the absolute configuration assigned[94, 96]. The C-24 stereochemistry of the 24-OH-D_3 epimers has also been determined[111, 345]. Recently, X-ray crystallographic analysis of the 25,26-diols was used to establish the absolute stereochemistry at C-25[331].

As possible modulators of vitamin D metabolism, several nitrogen-containing side chain analogs have been synthesized from the 22-aldehyde. The 25-aza vitamin [scheme (11)][346] has been found to be an inhibitor of D activity, while the 24-aza [scheme (12)] and the 24-amino vitamin [scheme (13)] derivatives (J. Chu, H. K. Schnoes, and H. F. DeLuca, unpublished results) have yet to be biologically tested.

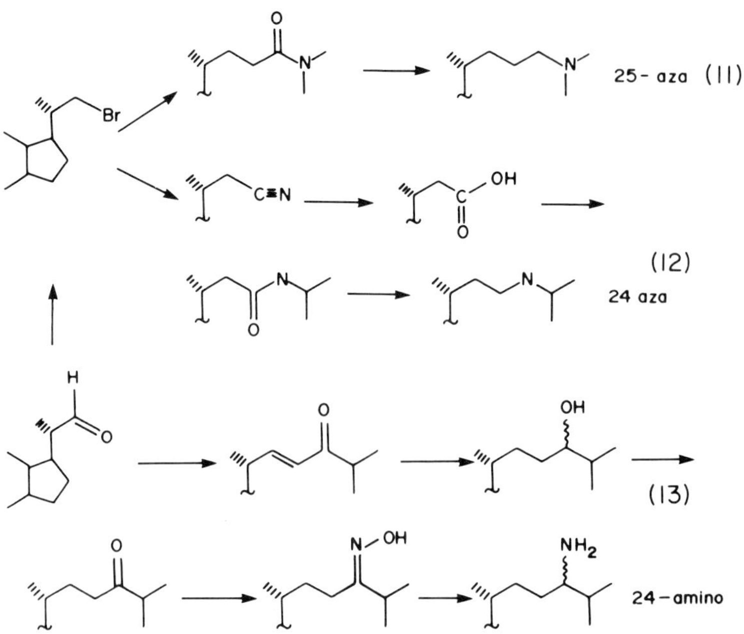

3 Steroidal A-Ring Functionalization

Introduction of oxygen at C-1 in the A-ring of the steroid nucleus usually is accomplished by basic α-epoxidation of the corresponding 3-keto-1-ene derivative. The scheme of Narwid et al. [scheme (14)][347] illustrates this approach in which the

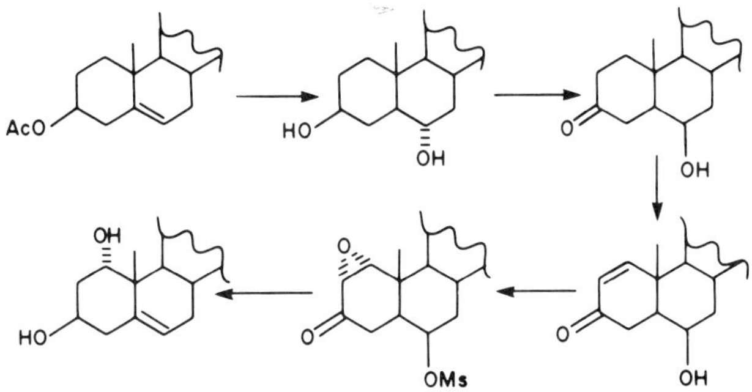

(14)

5,6-double bond is protected by conversion to 6-hydroxy or keto functions. Reduction of the epoxide and regeneration of the 5-ene gives the 1α-hydroxylated product in ca. 18% yield.

The approach of Hesse and Barton[348] [scheme (15)] to the 1α-hydroxy-5-cholestene system involves $Li/NH_3/NH_4Cl$ reduction of the 1α,2α-epoxy-4,6-dien-3-

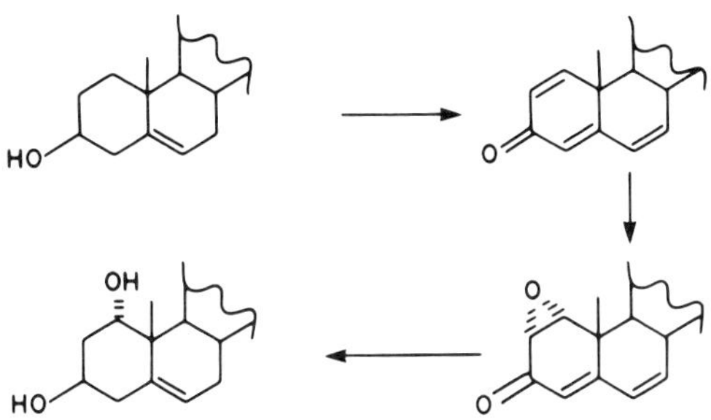

(15)

one generated from cholesterol by dehydrogenation with dichlorodicyanoquinone (DDQ) and basic epoxidation. The yield for this 3-step process is ca. 25% and it has been extensively used in the small-scale preparation of a variety of 1-hydroxylated vitamins[349–355].

A third approach [scheme (16)][356, 357] involves deconjugation of a 1,4-dien-3-one to the 1,5-diene followed by selective hydroboration of the 1,2-olefin. This scheme gives both the 1α-hydroxy and 2α-hydroxy cholesterol derivatives. Application of this method to 1,4,6-cholestatrien-3-one leads directly to the 1α-hydroxy provitamin D skeleton[358].

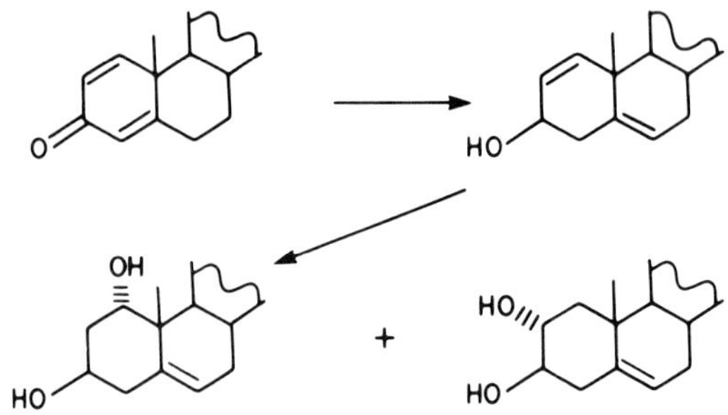

(16)

A fourth procedure [scheme (17)][359] is quite different from the previous examples. Oxygen is introduced at C-1 by the transannular photochemical 2+2 cycli-

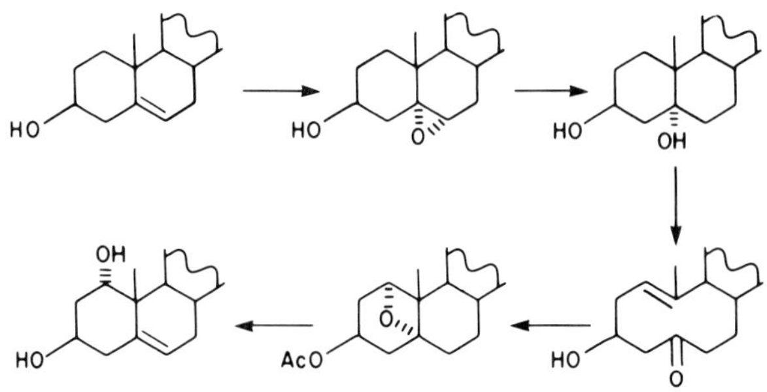

(17)

zation of the 5,10-seco steroid analog formed by HgO/hν oxidation of 5α-hydroxy cholesterol. The oxetane photoproduct is then opened in the presence of acid giving 1α-hydroxy-5-ene-steroid.

Epimerization reactions on steroidal A-ring alcohols have provided many analogs of potential biological interest.

The 3α-hydroxylated precursor to epi-D_3 has been synthesized via a stereoselective reduction of the 3-keto PTAD-5,7-diene adduct [scheme (18)][360] or direct

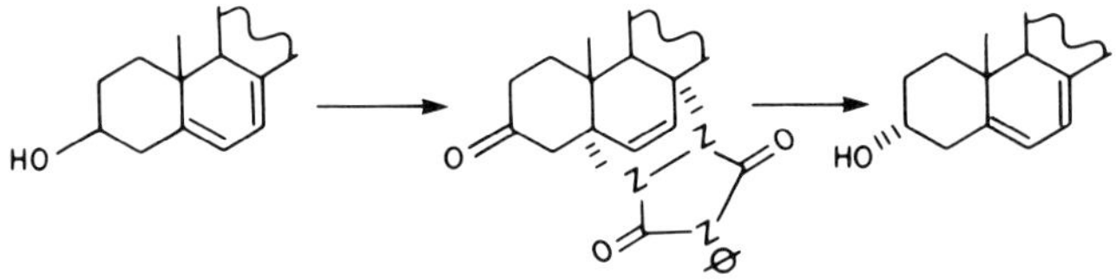

(18)

epimerization of the 3β-hydroxyl group with a system of triphenyl phosphine, diethylazodicarboxylate and formic acid [scheme (19)][361].

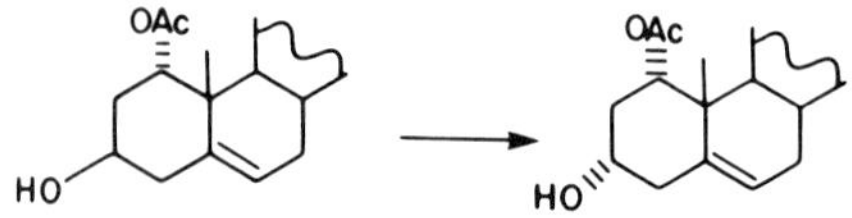

(19)

The 1β,3β-dihydroxyprovitamin was produced from 1-ketocholesterol acetate; however, this 1-hydroxy epimer failed to undergo photochemical ring opening to the previtamin analog (H. E. Paaren, H. K. Schnoes, and H. F. DeLuca, unpublished results).

In order to assess the importance of the 3-hydroxyl group with respect to vitamin D activity, 3-deoxy-1-hydroxylated analogs have been synthesized from steroidal precursor [scheme (20)][362, 363], and by the total synthetic approach of Okamura

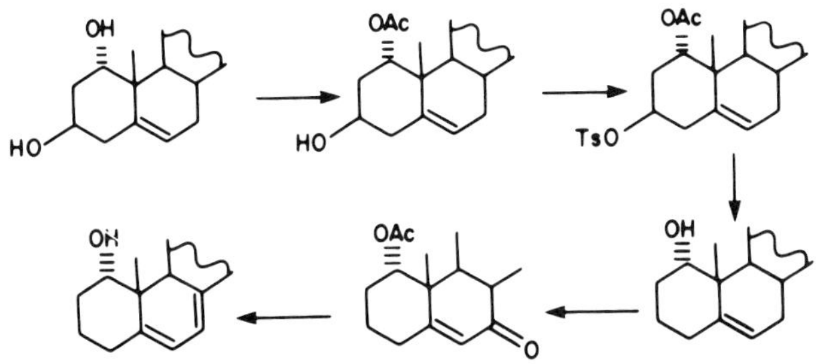

(20)

et al. [scheme (21)][364] in which the vitamin D triene system is formed by a unique thermal rearrangement of a vinyl-allene intermediate[365].

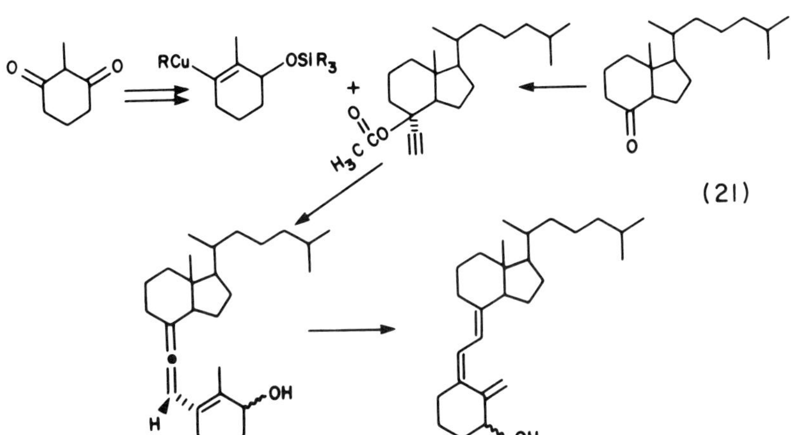

(21)

Although the partial synthesis of vitamin D was accomplished by Inhoffen nearly twenty years ago[366], most of the advances in the total synthesis of vitamin D have come from Lythgoe's laboratory. The total synthesis of 1α-hydroxyvitamin D_3 [scheme (22)][367] involves the general synthetic concept of coupling independently prepared ring A and ring C/D fragments. Yields are on the order of 20% from the starting chloroketone.

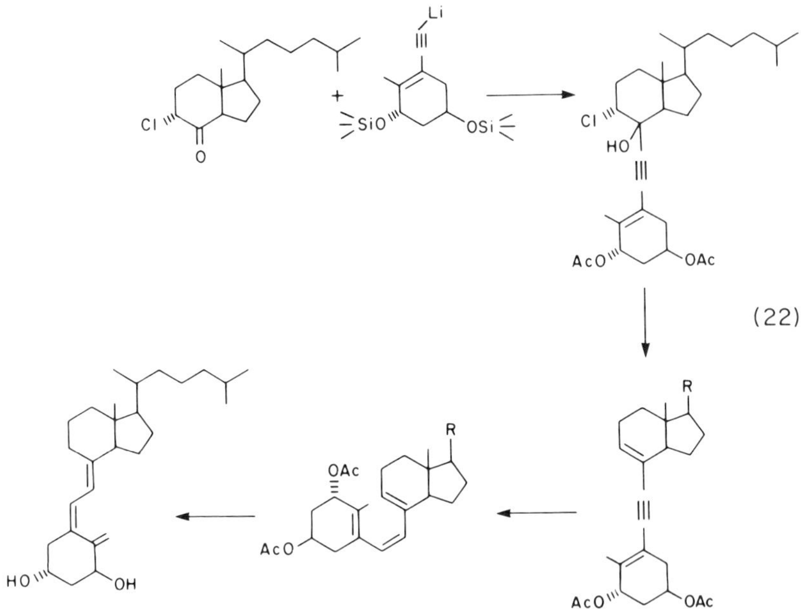

(22)

4 Synthetic Modification of Vitamin D

A different aspect of vitamin D chemistry involves direct synthetic manipulation of the vitamin triene skeleton. The earliest reactions of this type consisted of "dissolving metal reductions" conducted on vitamin D_3[368]. A series of dihydro derivatives of ill-defined stereochemistry were produced [scheme (23)]. The intrinsic re-

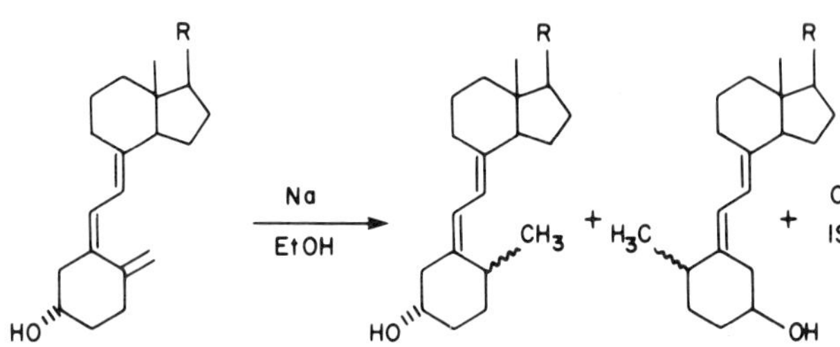

(23)

activity of the triene and complicated reaction mixtures seemed to dissuade early researchers from attempting further synthetic ventures along these lines. However, with recent advances in analytical and chromotographic techniques, research in this area has increased.

Much of the stereochemical uncertainties associated with the configuration of dihydrovitamin D and dihydrotachysterol have been resolved by recent studies[369]. By reacting *cis*- and *trans*-vitamin D_2 with 9-borobicyclononane followed by a basic peroxide workup, the stereoisomeric 10*S* and 10*R* 19-hydroxymethyl dihydro-derivatives were obtained in good yields. Tosylation of the 19-alcohol produced the intermediates which were either treated with a strong base (internal ether formation occurring *only* with the 10*R cis* isomer) or reduced to the 10,19-dihydro analogs [scheme (24)]. In this way the absolute stereochemistry at C(10) was established for many of the dihydro derivatives described in the older literature.

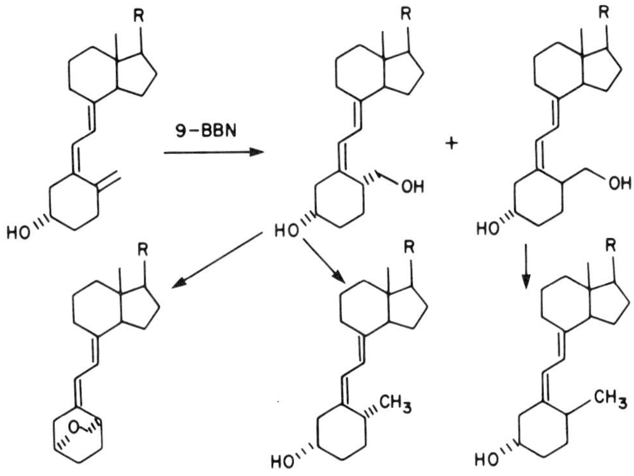

(24)

C(19)-hydroxy methyl dihydrovitamins have also been produced by $NaBH_4$ reduction of 19(E)-acetoxyvitamin D_3-acetate [scheme (25)][370].

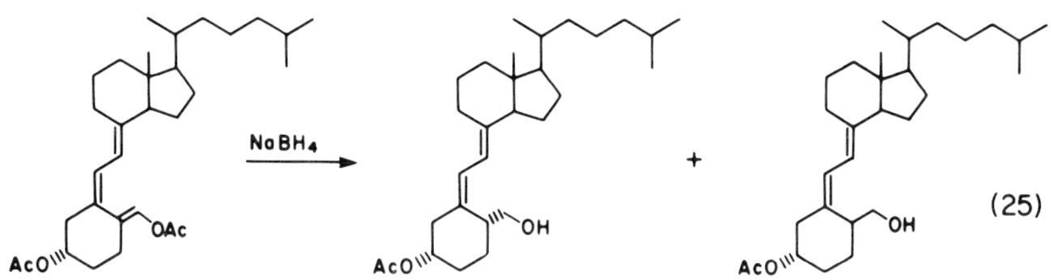

(25)

Many derivatives containing either fluorine or nitrogen have been synthesized directly from a vitamin-type structure.

The mild and selective reagent diethylamino sulfurtrifluoride (DAST), which introduces a fluorine atom for a hydroxyl group with retention of configuration, has been used to produce 25-fluorovitamin D_3[346], 1α-hydroxy-25-fluorovitamin D_3

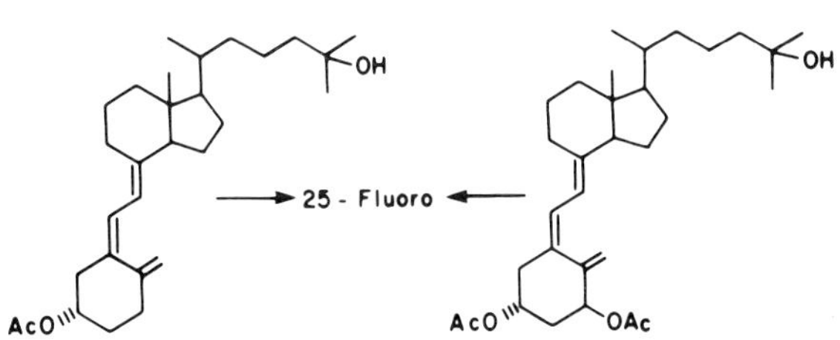

(26)

[scheme (26)][371] and 1α-fluorovitamin D_3 [scheme (27)][372]. All of these fluoro-vitamins possess interesting biological properties.

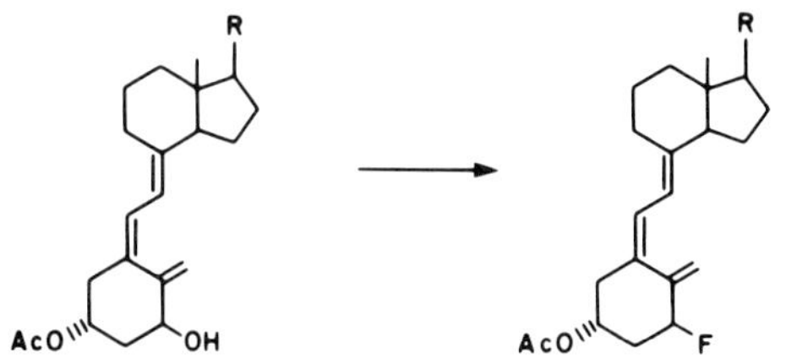

(27)

Fluorine has also been introduced stereospecifically at the 3β positions of 1α-OH-D_3 via hydrofluoric acid catalyzed cycloreversion of the 3,5-cyclo vitamin analog which was initially formed by solvolysis of the 3β-tosylate [scheme (28)][373].

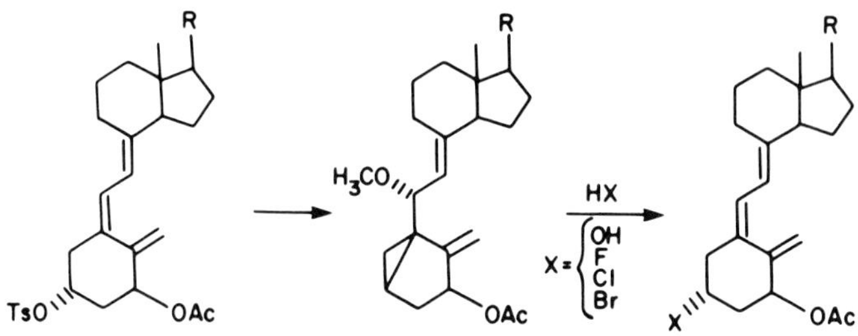

(28)

By using hydroazoic acid (HN_3) with triphenyl phosphine and ethylazodicarbo-xylate, vitamin D_3 and epi-vitamin D_3 were converted to the 3α and 3β azides respectively [scheme (29)]. Reduction with $LiAlH_4$ produced the amino analogs[374].

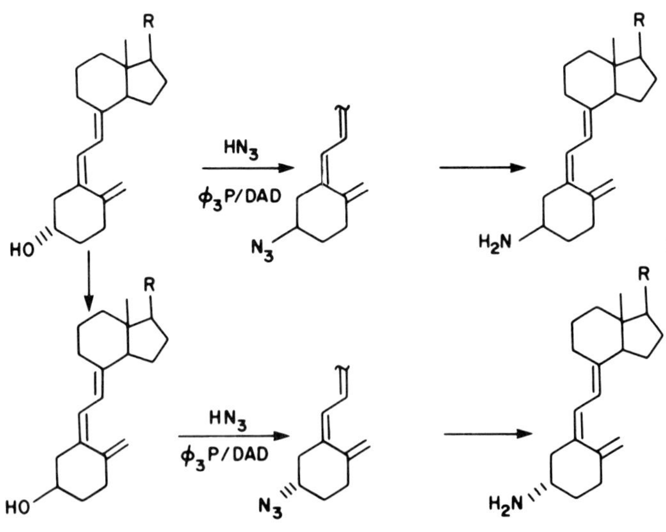

(29)

Epimerization of the 1α-hydroxyl group of 1α-OH-D_3 has been accomplished in a three-step sequence[56, 375]: MnO_2 allylic oxidation to the 1-keto previtamin analog, reduction to a mixture of 1α- and 1β-hydroxy previtamins, and thermal isomerization to the vitamin isomers [scheme (30)]. Vitamin D 1,3-diol stereoisomers

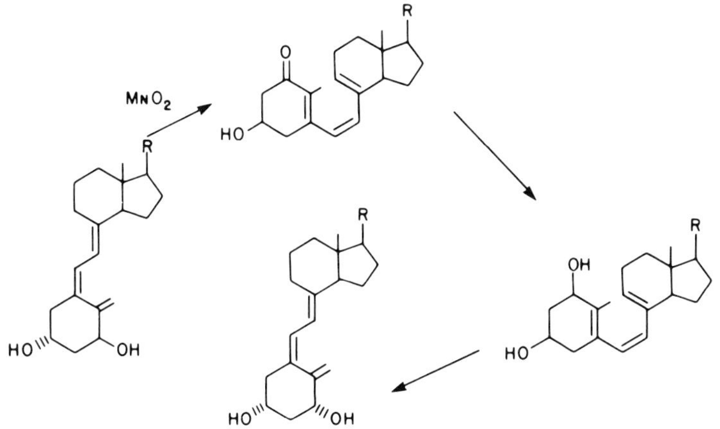

(30)

are useful in determining receptor binding priorities, and, therefore, aid in establishing structure-activity relationships.

From a practical viewpoint, direct C-1 hydroxylation of vitamin D_3 is a highly desirable reaction. However, direct allylic oxidation of vitamin D_3, for example, with selenium dioxide/t-butylhydroperoxide [scheme (31)][376, 377], produces only small amounts (< 5%) of 1-hydroxylated compounds.

These difficulties have been circumvented by using the 3,5-cyclovitamin as substrate for allylic oxidation [scheme (32)][377]. 1α-hydroxycyclovitamin D_3 is produced in ca. 50% yields and is readily converted to 1α-OH-D_3 by acid catalyzed solvolysis. The overall yield from vitamin D_3 ranges from 15–20%.

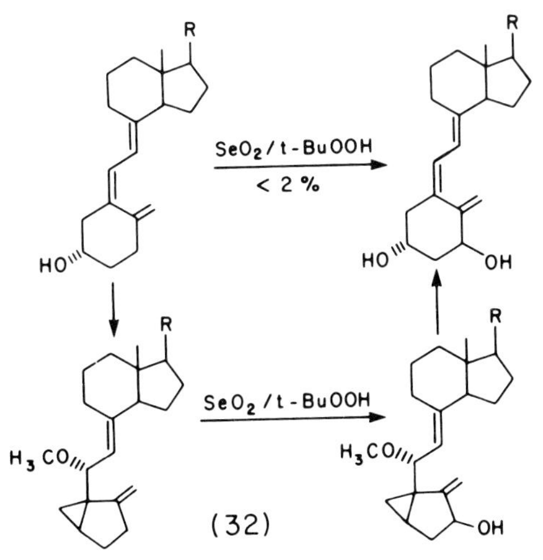

(31)

(32)

The entire set of stereoisomeric-1,3-dihydroxy *cis/trans* vitamin analogs, totalling eight in number, has recently been synthesized from vitamin D_3 and epi-vitamin D_3 via cyclovitamin intermediates and MnO_2- allylic hydroxyl epimerization reactions [scheme (33)]. (H. E. Paaren, H. K. Schnoes, and H. F. DeLuca, unpublished results).

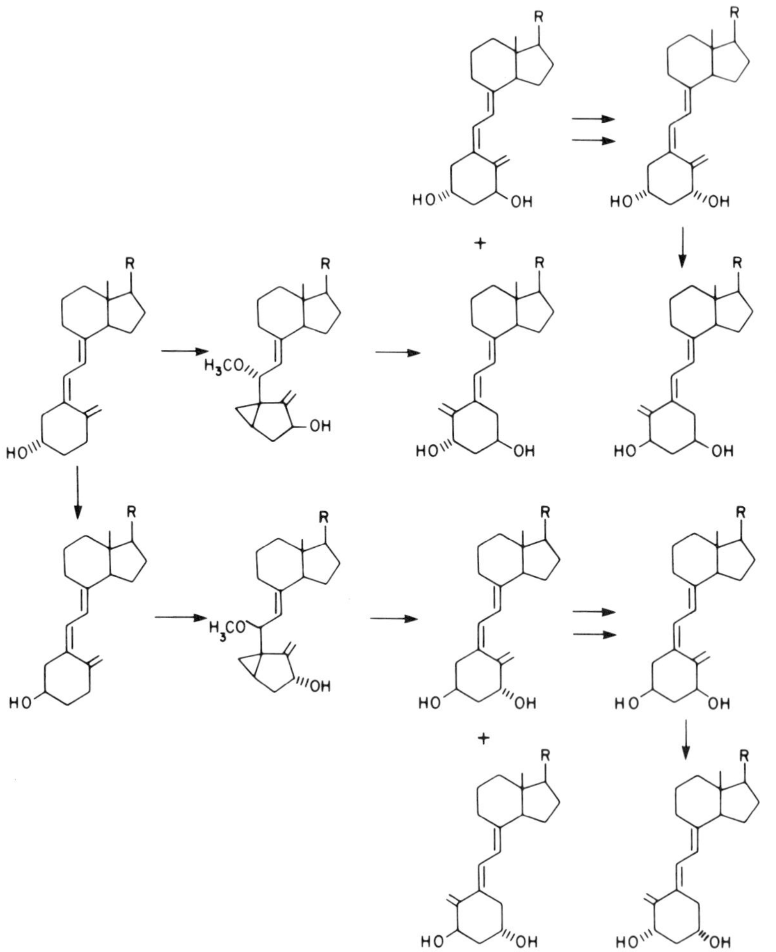

(33)

5 Radiolabeled Analogs

The recent advances in vitamin D metabolism and mode of action can be directly attributed to the synthesis of radiolabeled vitamin D derivatives of high specific activity.

Original syntheses employed catalytic reduction of an appropriate 1,2-steroidal olefin with tritium gas and produced ^{3}H-vitamin D_3 of moderate specific activity (5–10 Ci/mmole)[2, 3, 378]. Analogs of higher specific activity (20–80 Ci/mmole) can be obtained by catalytically reducing an acetylenic group in the side chain with T_2 [scheme (34)][283, 379] or by reacting the 25-carboxymethyl derivatives with CT_3MgBr [scheme (35)][50].

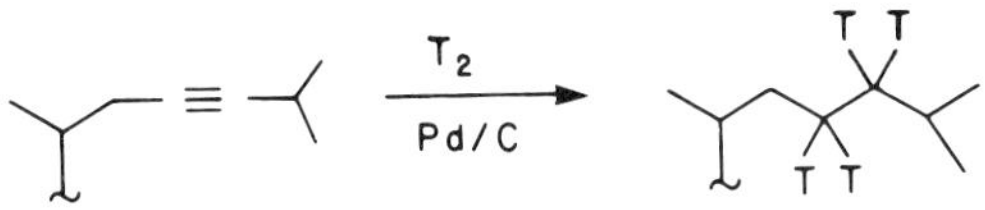

(34)

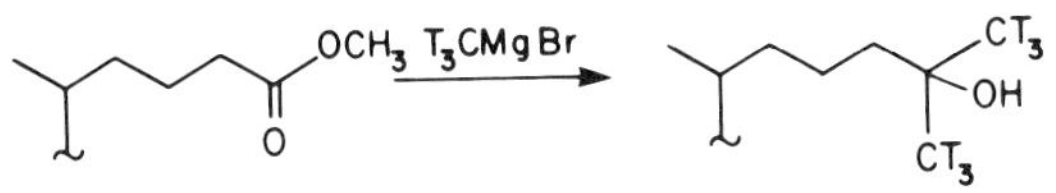

(35)

Another convenient synthesis of radiolabeled vitamin D has involved exchange in T_2O and reduction with $NaBT_4$ of the 3-keto vitamin analog in which the triene has been reversibly protected as an iron carbonyl adduct [scheme (36)] (S. Yamada, H. K. Schnoes, and H. F. DeLuca, unpublished results).

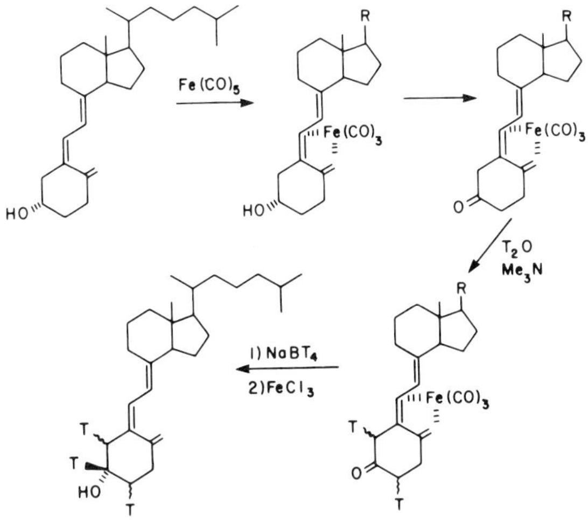

(36)

Even though synthetic vitamin D chemistry had its origins a half century ago, only now is the field beginning to develop with the innovative character apparent in other areas of organic chemistry. The organic chemist confronted with work in this area should always look forward to integrating new techniques and concepts in this well-established field.

VIII Structure-Activity Relationships

The availability of synthetic vitamin D metabolites and of analogs systematically modified at specific sites has made possible a rough definition of the relative importance of specific structural units for expression of biological activity. Comparisons between structural variants has led, for example, to a fairly clear assessment of the functional significance of the hydroxylation pattern, which establishes, not surprisingly, the key importance of the hydroxy groups introduced during vitamin D metabolism (at C-1α- and C-25). Similarly, the consequences of other alterations of the side chain and the triene chromophore and of stereo-chemical factors have been evaluated, and available data are sufficiently consistent to allow predictions on structure/activity relationships that should prove of some value for the design of analogs with specific properties. Both *in vivo* and *in vitro* assays have been used for activity comparisons between various vitamin D metabolites and analogs. *In vivo* assays include the determination of intestinal calcium transport, bone calcium mobilization and antirachitic activity, whereas the two most common *in vitro* systems exploited in recent years involve the measurement of calcium resorption from fetal rat bone in culture, and the determination of the relative binding affinity of a vitamin D ligand for the cytosolic protein from chick intestine thought to be the natural "receptor" for 1α,25-$(OH)_2D_3$. Although there are some gaps in the existing data, chiefly because few compounds have been tested in all systems, enough information is available to identify major relationships and trends. Especially the two *in vitro assays* mentioned offer a fairly clear picture of the effect of structural variations on activity, which allows a reasonably consistent qualitative ranking of distinct structural parameters.

1 Structural Requirements for Receptor Binding

As discussed in a preceeding Section VI.1, non-covalent binding of 1α,25-$(OH)_2D_3$ to a cytosolic receptor protein in intestine is thought to be an essential preliminary step for expression of biological activity, i. e. the activation of the intestinal calcium transport system. A comparative study of the affinity of selected structural analogs for the receptor thus provides a direct and fairly quantitative assessment of the relative importance of specific vitamin structural units for ligand/receptor interaction, and by inference, therefore, of their effect on biological potency. The groups of Haussler[293, 395], Norman[396, 397], and DeLuca[81, 82] have investigated the binding between various vitamin D compounds and the intestinal receptor system. The assay procedure of the first two laboratories involved a combined cytosol receptor/nuclear chromatin system and their measurements therefore reflect the combination of vitamin/receptor and receptor complex/chromatin interactions. The assay of DeLuca and co-workers is more defined in that the chromatin binding step is eliminated and only the initial binding reaction between ligand and proteins receptor is under observation, but both assay methods give comparable results, as far as structure/activity considerations are concerned.

Available data[82, 396] on receptor/vitamin D interaction support the general conclusion that affinity for the receptor is determined primarily by the spatial arrangement of the hydroxy functions of the ligand, and that among the three hydroxy groups of 1α,25-$(OH)_2D_3$, those at C-1 and C-25 exert the most profound effect. The data shown in Fig. 11 offer dramatic confirmation of this general statement: 1α,25-$(OH)_2D_3$ (13g), the active metabolite in intestine, exhibits highest

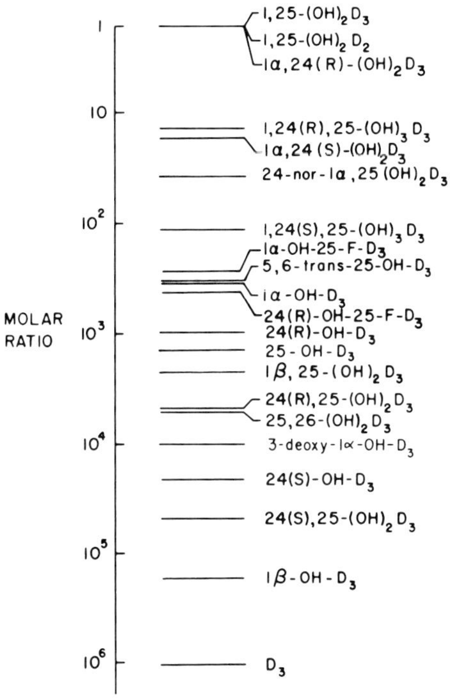

Fig. 11. Relative ability of vitamin D metabolites and structural analogs to compete with 1,25-$(OH)_2$-[23,24-^{3}H]D_3 for the receptor for 1,25-$(OH)_2D_3$ in chick intestinal cytosol. The log of the ratio of moles/liter of analog over the moles/liter of 1,25-$(OH)_2$-[^{3}H]D_3 required for 50% displacement of the ^{3}H from the receptor is plotted

affinity for its receptor and the absence of 1- and 25-hydroxy functions (as in vitamin D_3 itself) reduces binding affinity by six orders of magnitude[82]. The addition of a 25-hydroxy group or a 1-hydroxy group to vitamin D_3 improves affinity 1000- and ca. 5000-fold, respectively (see values of 25-OH-D_3 and 1α-OH-D_3 in Fig. 11), demonstrating the key importance of these two functions for receptor/ligand interaction. In contrast, the C-3-hydroxy group, has a more modest effect. The absence of a 3-hydroxy function, as in 3-deoxy-1α,25-$(OH)_2D_3$ (17g), for example, reduces binding by only a factor of 10 compared to 1α,25-$(OH)_2D_3$[396] and the data of Fig. 11 demonstrate that the analogous pair –– 1α-OH-D_3 (13a) vs. 3-deoxy-1α-OH-D_3 (17a) –– exhibit roughly the same relationship in terms of binding affinity[82]. A noteworthy result is the demonstration that the binding interaction appears only moderately sensitive to the exact position of the side chain hydroxy group. Thus the binding of analog 24(*R*)-OH-D_3 (1b) is essentially equal to that of 25-OH-D_3 (1g) and 1α,24(*R*)-$(OH)_2D_3$ (13b) is equivalent to 1α,25-$(OH)_2D_3$ in terms of binding affinity (see Fig. 11). The binding site of the receptor thus appears to accommodate a 24- or 25-OH function with roughly equal facility, and that a 24-hydroxy group provides for strong receptor/ligand interactions is further confirmed by the more recent observation that 24-nor-1α,25-$(OH)_2D_3$ (13n) competes quite effectively (50-fold excess) with 1α,25-$(OH)_2D_3$ for receptor sites and that 24-hydroxy-25-fluorovitamin D_3 exhibits an affinity comparable to that of 25-OH-D_3 (see Fig. 11).

In the presence of a 25-hydroxy group, the introduction of an additional hydroxy group into the side chain at either C-24 or C-26 leads to a significant reduction in binding interaction. The relative binding affinities of 1α,25-$(OH)_2D_3$ vs. 1α,24(*R*),25-$(OH)_3D_3$ (13h), and of 25-OH-D_3 vs. 24(*R*),25-$(OH)_2D_3$ (1h) or 25,26-$(OH)_2D_3$ (1j) (Fig. 11) provide clear evidence of this effect[82], and a comparison between the respective 24(*R*) and (*S*) stereoisomers listed in Fig. 11 further demonstrates that the stereochemistry of the hydroxy group at C-24 is of consequence; in each case, the unnatural 24(*S*)-isomer is the less effective competitor by roughly a factor of 10. The reason for this diminished binding affinity attendent upon the introduction of a C-24-hydroxy group is not understood. An explanation involving electronic effects –– for example, a reduction in the strength of hydrogen-bonding between the C-25-hydroxyl and the receptor caused by internal H-bonding between C-24 and C-25-hydroxy groups –– is perhaps most plausible, since an argument based on a purely steric effect of the C-24 substituent is weakened by the observation that a C-24 methyl substituent (as in 1α,25-$(OH)_2D_2$ (13d) –– where the 24-methyl group occupies the position of the 24(*S*)-hydroxy function in 1,24(*S*),25-$(OH)_3D_3$ (13i) or 24(*S*),25-$(OH)_2D_3$ (1i) has no effect on receptor/ligand interaction (see Fig. 11).

Given the importance of the 1-hydroxy group in the binding interaction it is not surprising that the stereochemistry of this function has a marked effect on binding. Thus 1β,25-$(OH)_2D_3$ (16g) is ca. 3000 times, and 1β-OH-D_3 (16a) is 10^5–10^6 times less effective than 1α,25-$(OH)_2D_3$ in competing for receptor sites[56) –– an affinity roughly comparable to the corresponding C-1-deoxy compounds cited in Fig. 11. The effect of 3-hydroxy stereochemistry on binding has not yet been investigated, although suitable models, e. g. 1α-OH-3-epi-D_3 (16a) are available[360, 361].

The importance of the 1-hydroxy group is also exemplified by the relatively good binding interaction of certain triene-modified derivatives. For example, 5,6-*trans*-25-OH-D_3 [(5g) see Fig. 11] and 25-OH-DHT_3 (10g)[396] (~500 and ~ 100 times, respectively, less active than 1α,25-$(OH)_2D_3$), are effective ligands, presumably by virtue of the 180 °-rotation of ring A which transposes the C-3β-hydroxy function into a pseudo-1α-hydroxy substituent. The fifty-fold difference in binding of 3-deoxy-1α,25-$(OH)_2D_3$ (17a) (ca. 1/10 of 1α,25-$(OH)_2D_3$) and 5,6-*trans*-25-OH-D_3 (ca. 1/500 relative to 1α,25-$(OH)_2D_3$) implies perhaps an unfavorable steric interaction between the transposed exocyclic methylene unit of the latter and the receptor binding cavity, which reduction to methyl (giving 25-OH-DHT_3 (10g) ca. 1/100 relative to 1α,25-$(OH)_2D_3$) appears to alleviate measurably.

The preceding discussion leads to the fairly obvious conclusion that most structural departures from 1α,25-$(OH)_2D_3$ entail "penalties" in terms of ligand/receptor affinity. Deletion of C-1 or C-25-OH functions, or change of hydroxy stereochemistry at C-1 causes pronounced perturbations (factors of ca. 500–1000), introduction of additional hydroxy groups at C-24 or C-26, the deletion of the C-3-hydroxy function, the stereochemistry at C-24, the transposition of ring A results in more subtle modulations (factors of ca. 10–50). That a 24(*R*)-hydroxy group is essentially equivalent to a C-25-OH-function is an important finding and of note is also the observation that one, at first sight, rather considerable change –– replacing the D_3-side chain with that of vitamin D_2 –– has no measurable effect, suggesting that exact side chain structure may be of secondary consequence provided the appropriate length in preserved –– to allow for introduction of the key hydroxy group. It is worth mentioning in this connection that elimination of most of the side chain, as in 1α-hydroxypregnacalciferol (1α-OH-PC (13w), see Fig. 10) results in an analog with no *in vivo* activity[353], but surprisingly high affinity for the 1α,25-$(OH)_2D_3$-receptor (1/70 of that of 1α,25-$(OH)_2D_3$) (W. S. Mellon, H. K. Schnoes, and H. F., DeLuca, unpublished). This is a somewhat puzzling result, in light of the foregoing discussion, but one that deserves to be further explored with suitably designed derivatives.

2 In Vitro Bone Resorption

The bone resorption assay measures the release of calcium from fetal rat bone in culture in response to addition of vitamin D compounds. Although the assay, therefore, relates to a different organ and monitors a decidedly more complex process than that involved in the protein binding assay discussed above, the relative order of activity exhibited by the various vitamin metabolites and analogs that have been tested in both systems is remarkably similar. Thus, 1α,25-$(OH)_2D_3$ is again the most potent compound, eliciting a pronounced bone resorption response in the concentration range of 10^{-11}–10^{-10}M[314, 315, 398, 399]. Removal of either the C-25 or C-1 hydroxy group [i. e. 1α-OH-D_3[315, 353, 398] or 25-OH-D_3[76, 77]] has a fairly dramatic effect, reducing activity 500–1000-fold, whereas the elimination of the 3β-OH-function (as in 3-deoxy-1α,25-$(OH)_2D_3$ (17a)) diminishes potency by only a factor of ca. 50[400]. A change of hydroxy substitution from C-25 to C-24 affects activity only moderately since 24(*R*)-OH-D_3 is roughly as effective as 25-OH-D_3[316], and the potency of 24-nor-1α,25-$(OH)_2D_3$ is only 50 times lower than that of 1α,25-$(OH)_2D_3$[401]. These comparisons imply that analogs such as 1α,24(*R*)-$(OH)_2D_3$ should approximate 1α,25-$(OH)_2D_3$ in potency. C-24-hydroxy stereochemistry again has a significant effect: 24(*R*)-OH-D_3 (1b) is at least 10 times more active than its (*S*)-isomer (1c)[316], and the same relationship holds for the pairs 24(*R*),25-$(OH)_2D_3$ (1h) and 24(*S*),25-$(OH)_2D_3$ (1i)[316], and 1α,24(*R*),25-$(OH)_3D_3$ (13h) and 1α,24(*S*),25-$(OH)_3D_3$ (13i)[399]. As was observed in the intestinal receptor assay, introduction of a 24-hydroxy group (into a 25-hydroxy analog) diminishes activity: the relative potency of the pairs, 25-OH-D_3 vs 24(*R*),25-$(OH)_2D_3$ and 1α,25-$(OH)_2D_3$ vs. 1α,24(*R*),25-$(OH)_3D_3$ differs rather consistently by a factor of ca. 10[316, 399], and, as just mentioned, the stereochemistry of the 24-substituent (i. e. change from R to S) then further modulates expression of activity[316]. The reduction of activity is peculiar to the 24-hydroxy-substituent, since 1α,25-$(OH)_2D_2$ (13d) –– bearing a 24-methyl group –– is as effective as 1α,25-$(OH)_2D_3$[399]. Little information is available on the effect of other structural modifications. Analogs such as 25-OH-DHT_3[402] and 5,6-*trans*-25-OH-D_3[398] –– which lack a "C-3"-hydroxy function and feature also a modified triene system –– exhibit potencies similar to those of 1α-OH-D_3 and 25-OH-D_3. A comparison between 3-deoxy-1α,25-$(OH)_2D_3$ and 5,6-*trans*-25-OH-D_3, again suggests that reorientation of ring A by 180 °, accounts for a ~ 10-fold decrease in bone resorption activity. All compounds carrying only a single hydroxy group, e. g. D_3, DHT_3 (10a), 5,6-*trans*-D_3 (5a) and 3-deoxy-1α-OH-D_3 (17a), which are active *in vivo* because of metabolism to their

25-hydroxy or 1α,25-dihydroxy forms — show no activity in the bone resorption assay, presumably because even the highest experimentally feasible concentration (10^{-5}–10^{-4}M, because of solubility limitations) do not suffice for expression of the low intrinsic activity of these compounds.

3 Activity in Vivo

Whereas the *in vitro* assays discussed here give a fairly clear-cut and consistent order of relative potency among analogs, comparative activity assessments *in vivo* are complicated by a variety of factors, the influence and magnitudes of which usually cannot be gauged very accurately. Metabolism of analogs to more (or less) active derivatives, rates of metabolism, unknown tissue distribution, rates of excretion, and differences in time-course of response all may affect eventual expression of activity. The *in vivo* activity estimate represents an integration of all these factors, and to find differences between the *in vitro* and *in vivo* responses elicited by certain analogs is hardly surprising. Most of these differences can be rationalized by assuming the intervention of metabolic conversion steps *in vivo*.

Assessments of *in vivo* activity are based primarily on two assays: the stimulation of intestinal calcium transport and the mobilization of calcium and phosphate from bone in animals (rats or chicks) maintained on a low calcium or low phosphorus diet. Many of the relevant results in this area have been reviewed previously[403, 404], and are also mentioned in preceding sections of this review (Section III); we can restrict discussion here then to a summary of general trends and conclusions. *In vivo* and *in vitro* results agree as to the key importance of the hydroxy functions at C-1 and C-25; 1α,25-$(OH)_2D_3$, is the most potent and fastest acting compound *in vivo*[98, 121, 405], and at physiological concentrations, C-1-deoxy compounds such as 25-OH-D_3 or 24,25-$(OH)_2D_3$ are inactive when 1-hydroxylation is prevented by nephrectomy whereas the corresponding 1-hydroxy forms are fully potent unter these conditions[70, 72, 73, 93, 101]. The corresponding vitamin D_2 analog, 1α,25-$(OH)_2D_3$, is identical to 1α,25-$(OH)_2D_3$ in mammals[121], but as discussed previously is substantially (5–10X) less active in birds[114]. The importance of a 25-hydroxy function is equally evident, even though certain 25-deoxy analogs — such as 1α-OH-D_3[356, 391, 406, 407], 1α-OH-D_2 (13d)[351], or 3-deoxy-1α-OH-D_3[363] — exhibit very pronounced potency *in vivo.* For the case of 1α-OH-D_3 it can be shown, however, that expression of activity is preceded by metabolism, in the rat[408, 409] and in the chick[61], to 1α,25-$(OH)_2D_3$ and this *in vivo* conversion to the natural hormone also explains the marked discrepancy between the apparent *in vitro* activity of the substance (i.e. 500 times lower than 1α,25-$(OH)_2D_3$ in receptor or the bone resorption assays) and its *in vivo* efficacy (ca. nearly as effective as 1α,25-$(OH)_2D_3$ in rats and chicks)[406, 407]. The demonstration that other 25-deoxy analogs — such as DHT_3 or 5,6-*trans*-D_3 — are also metabolized to the corresponding 25-hydroxyated forms[410, 411] explains their high potency *in vivo*[393], even though these compounds are essentially inactive in the two *in vitro* assays discussed earlier.

In vivo results also are in fair agreement with *in vitro* data as to the importance of the exact position of the side chain hydroxy function. For example, 24-nor-1α,25-

$(OH)_2D_3$ is estimated to be ca. 100 times less active than $1\alpha,25\text{-}(OH)_2D_3$ in stimulating intestinal calcium transport[401], a number in reasonable accord with its ca. 50-fold lower affinity for the intestinal receptor protein. Similarly, the analog 1α-hydroxy-25-fluorovitamin D_3 (13t), in which C-25 is blocked to hydroxylation[371], exhibits a remarkably high *in vivo* activity –– 1/50 that of $1\alpha,25\text{-}(OH)_2D_3$ –– although in its affinity for the intestinal receptor it mimics $1\alpha\text{-}OH\text{-}D_3$ (i. e. ca ~ 400 lower than $1\alpha,25\text{-}(OH)_2D_3$). This difference implies *in vivo* metabolism to a more active form –– very likely the 24-hydroxylated derivative –– 1α,24-dihydroxy-25-fluorovitamin D_3. Metabolism to a 1α-hydroxylated form also offers a plausible explanation for the observation that the 24-hydroxy-25-fluorovitamin D_3 is as active as $25\text{-}OH\text{-}D_3$ in all systems tested[412]. *In vivo* and *in vitro* results thus agree on the effect of side chain hydroxyl substitution: full activity requires a C-25-hydroxy function, but a 24-hydroxyl is a very effective substitute which entails perhaps some reduction but not total abolition of activity. When side chain hydroxylation is not possible, however, as in the case of 1α-hydroxypregnacalciferol (1α-OH-PC, see Fig. 10 for structure) an inactive analog results[353]. Expression of activity by 24(*R*)- and 24(*S*)-$OH\text{-}D_3$ appears to require a two-step transformation to the corresponding 1α,25-hydroxylated forms[113]. Of particular interest is the observation that both compounds are readily converted to the corresponding 24,25-dihydroxylated forms, but only the (*R*)-epimer undergoes subsequent C-1-hydroxylation to any appreciable extent to yield $1\alpha,24(R),25\text{-}(OH)_3D_3$, the presumed active form[113]. Since the (*R*) and (*S*) forms of $1\alpha,24,25\text{-}(OH)_3D_3$ are essentially equal in potency *in vivo*[104], the discrimination of the 1-hydroxylation step in kidney explains the notable activity difference[112] between the (*R*) and (*S*) epimeric pairs of $24\text{-}OH\text{-}D_3$ and 24,25-$(OH)_2D_3$.

The effect of ring A-hydroxy stereochemistry on *in vivo* activity has not yet been investigated in great detail. The $1\beta\text{-}OH\text{-}D_3$ analog is reported to be inactive[419], a finding in accord with the very low affinity of this compound for the receptor protein in intestine[56]. The absence of the 3β-OH-function reduces *in vivo* activity, but not dramatically: 3-deoxy-$1\alpha,25\text{-}(OH)_2D_3$ is estimated to approximate the potency of $1\alpha,25\text{-}(OH)_2D_3$[397] and 3-deoxy-$1\alpha\text{-}OH\text{-}D_3$ is 20–50 times less active than $1\alpha\text{-}OH\text{-}D_3$, a result that again implies metabolic conversion (to 3-deoxy-1α,25-$(OH)_2D_3$) since 3-deoxy-$1\alpha\text{-}OH\text{-}D_3$ itself exhibits either low (receptor binding) or no (bone resorption) activity *in vitro*[363].

When the possible metabolic conversions and the inherent uncertainties in the biological activity estimates are taken into account, there is a fairly good correlation between the data contributed by different assays. Certainly *in vivo* and *in vitro* agree on the key activity-conferring structural parameters, and secondary structural effects are also reflected reasonably faithfully by the available data. Correlation breakdowns (i. e. the considerable activity difference between vitamin D_3 and its metabolites vs. vitamin D_2 and its metabolites in birds) are more of an exception, resulting in this specific case from an apparent species-specific degradative metabolism[115] that appears to be triggered by a C-24-substituent, since 24-substituted D_3 analogs, i. e. all 24-hydroxylated forms, are also relatively less effective in the chick than in the rat (Section IV).

4 Concluding Summary

The two *in vitro* assays discussed above — vitamin binding to intestinal receptor and vitamin-induced calcium resorption from bone — leave little doubt that, given an intact vitamin D skeleton the hydroxy groups of the molecule are primary determinants of activity. In both assays, $1\alpha,25\text{-}(OH)_2D_3$ is the most active compound known. Elimination of either the C-1 or C-25-hydroxy function has a dramatic effect, reducing affinity for the receptor or calcium resorption ability by 2–3 orders of magnitude, wheras deletion of the 3-hydroxy group leads to only a roughly 10-fold reduction of activity. A shift of the side chain hydroxyl from C-25 to C-24 has hardly any effect on binding affinity or bone resorption, if the 24-hydroxy group is in the (*R*)-configuration (i. e. the natural stereochemistry) but the 24(*S*)-epimer is ten times less active. Similarly the introduction of a 24(*R*)-hydroxy group into a 25-hydroxylated vitamin analog diminishes potency ten-fold, and the activity of the corresponding 24(*S*),25-dihydroxylated analogs is another 0.5–1 order of magnitude lower. Introduction of a 26-hydroxy function into 25-OH-D_3, decreases *in vitro* activities by a similar factor. The stereochemistry of the 1-hydroxy group is of even greater influence in view of the roughly thousand-fold difference in receptor binding affinity between $1\alpha,25\text{-}(OH)_2D_3$ and $1\beta,25\text{-}(OH)_2D_3$ and between 1α-OH-D_3 and its 1β-epimer. The orientation of the triene system has a significant but not dramatic effect: 5,6-*trans*-25-OH-D_3 is about 10 times less effective than 3-deoxy-1α,25-$(OH)_2D_3$ (compare structures in Fig. 10) in competing for receptor sites, and the same order of activity holds in the bone resorption assay.

Activity *in vivo* follows the *in vitro* pattern fairly consistently as long as allowance is made for the activating (or deactivating) effect of metabolic conversions. It is hydroxylation at C-25, for example, that explains the *in vivo* potency of such analogs as 5,6-*trans*-D_3, DHT_3 or 3-deoxy-1α-OH-D_3, which elicit no measurable or very low responses in the *in vitro* assays, and it is the formation of the 1α,25-dihydroxylated vitamin D compounds, by hydroxylation at C-1 or C-25, that accounts for the dramatic *in vivo* activity of 25-OH-D_3 or of 1α-OH-D_3 and 1α-OH-D_2. Similarly, the difference between the *in vivo* responses elicited by 24(*R*),25-$(OH)_2D_3$ and its 24(*S*) epimer appears to be a reflection primarily of the much more efficient 1α-hydroxylation of the former compound, modulated further by the intrinsically lower activity of the unnatural 24(*S*)-form in the target tissues.

No analogs with true organ specificity have been reported as yet, but some selectivity of action can be demonstrated. For example, the 24(*S*)-epimers of 24-OH-D_3 and 24,25-$(OH)_2D_3$ are reported to be relatively more effective *in vivo* in intestine than in bone, and a recently prepared compound, 1-fluoro-vitamin D_3 (J. L. Napoli, M. A.,Fivizzani, H. K.,Schnoes, and H. F. DeLuca, unpublished results), appears to have the opposite effect.

Finally, there is only one well-documented species difference in the intestinal and bone response to vitamin D compounds. All C-24-substituted vitamin D compounds, e. g. all D_2-derivatives and the 24-hydroxy-D_3 analogs, are significantly less effective in birds than in other animals. This difference is believed to be caused by the more rapid degradation and excretion of D_2 metabolites and 24-hydroxy-D_3

analogs in birds, since there is no discrimination in the required hydroxylation steps nor, apparently, at the site of action.

The information summarized in the preceding section provides a good basic framework for assessing the consequences of selected structural modification on expression of vitamin D activity. Available data from all assays are sufficiently consistent to serve as a useful and reasonably reliable guide for future synthetic efforts designed to produce compounds with specific biological properties, and the chemical work, in turn, can be expected to add many of the details currently missing from our picture of structure/activity relationships.

IX References

1. Kodicek, E., in: Proc. Fourth Internat. Congress of Biochemistry, Vol. XI, p. 198. Umbreit, W., Molitor, H. (eds.). London: Pergamon Press, Ltd. 1960
2. Norman, A. W., DeLuca, H. F.: Biochemistry *2,* 1160 (1963)
3. Neville, P. F., DeLuca, H. F.: Biochemistry *5,* 2201 (1966)
4. Norman, A. W., DeLuca H. F.: Anal. Chem. *35,* 1247 (1963)
5. Burlingame, A. L., Smith, D. H., Olsen, R. W.: Anal. Chem. *40,* 13 (1968)
6. Norman, A. W., Lund, J., DeLuca, H. F.: Arch. Biochem. Biophys. *108,* 12 (1964)
7. Lund, J., DeLuca, H. F.: J. Lipid Res. *7,* 739 (1966)
8. DeLuca, H. F.: Fed. Proc. *33,* 2211 (1974)
9. DeLuca, H. F., in: Urolithiasis research, p. 165. Smith, L. H., Vahlensieck, W. (eds). New York: Plenum Publishing Corp. 1976
10. Glisson, F., as cited in: Rickets, osteomalacia and tetany, (A. F. Hess). Philadelphia: Lea & Febiger 1929
11. Whistler, D., as cited in: Smerdon, G. T.: Daniel Whistler and the English Disease, a translation and biographical note. H. Hist. Med. *5,* 397 (1950)
12. DeLuca, H. F., in: The fat-soluble vitamins, Vol. II, Handbook of Lipid Research, p. 69. DeLuca, H. F. (ed). New York: Plenum Press 1978
13. Hess, A., in: Rickets, including osteomalacia and tetany, p. 22. Philadelphia: Lea & Febiger 1929
14. Eichmann, C.: Virchows Arch. Pathol. Anat. Physiol. *148,* 523 (1897)
15. Liebig, J. von: Ann. Chem. *39,* 129 (1841)
16. Hart, E. B., McCollum, E. V., Steenbock, H., and Humphrey, G. C.: Res. Bull. Univ. Wis. Agr. Expt. Station, June 17, 1911
17. Magendie, F.: Ann. de Chim et de Phys. *3,* 66 (1816), as cited in: McCollum, E. V.: A history of nutrition, p. 86. Glass, H. B., (ed.). Boston: Houghton-Mifflin Co. 1957
18. Grijns, G.: Geneesk. Tijdschr. V. Ned. *1,* (1901), as cited in: McCollum, E. V.: A history of nutrition, p. 216. Glass, H. B., (ed.). Boston: Houghton-Mifflin Co. 1957
19. McCollum, E. V., Davis, M.: J. Biol. Chem. *15,* 167 (1913)
20. Funk, C.: J. Physiol. *43,* 395 (1911)
21. McCollum, E. V., Simmonds, N., Pitz, W.: J. Biol. Chem. *27,* 33 (1916)
22. Mellanby, E.: Lancet *1,* 407 (1919)
23. Mc Collum, E. V., Simmonds, N., Becker, J. E., Shipley, P. G.: J. Biol. Chem. *53,* 293 (1922)
24. Huldshinsky, K.: Deut. Med. Wochschr. *45,* 712 (1919)
25. Chick, H., Palzell, E. J., Hume, E. M.: Medical Research Council, Special Resport No. 77 (1923)
26. Goldblatt, H., Soames, K. M.: Biochem. J. *17,* 446 (1923)

27. Steenbock, H.: Science *60,* 224 (1924)
28. Steenbock, H., Black, A.: J. Biol. Chem. *61,* 405 (1924)
29. Steenbock, H., Black, A.: J. Biol. Chem. *64,* 263 (1925)
30. Hess, A. F., Weinstock, M., Helman, F. D.: J. Biol. Chem. *63,* 305 (1925)
31. Askew, F. A., Bourdillon, R. B., Bruce, H. M., Jenkins, R. G. C., Webster, T. A.: Proc. Roy. Soc. *B107,* 76 (1931)
32. Windaus, A., Linsert, O., Lüttringhaus, A., Weidlich, G.: Ann. *492,* 226 (1932)
23. Bills, C. E., in: The vitamine, Vol. II. p. 164. Sebrell, W. H., Jr., Harris, R. S., (eds.). New York: Academic Press
34. Windaus, A., Lettre, H., Schenck, F.: Ann. *520,* 98 (1935)
35. Windaus, A., Schenk, F., von Werder, F.: Hoppe-Seylers Z. Physiol. Chem. *241,* 100 (1936)
36. Howland, J., Kramer, B.: Am. J. Diseases Children *22,* 105 (1921)
37. Shipley, P. G., Kramer, B., Howland, J.: Biochem. J. *20,* 379 (1926)
38. Orr, W. J., Holt, L. E., Jr., Wilkins, L., Boone, F. H.: Am. J. Diseases Children *26,* 362 (1923)
39. Nicolaysen, R.: Biochem. J. *31,* 122 (1937)
40. Nicolaysen, R.: Biochem. J. *31,* 107 (1937)
41. Nicolaysen, R.: Acta Physiol. Scand. *6,* 201 (1943)
42. Wasserman, R. H. (ed.): The transfer of calcium and strontium across biological membranes. New York: Academic Press 1963
43. Carlsson, A.: Acta Physiol. Scand. *26,* 212 (1952)
44. Nicolaysen, R., Eeg-Larsen, N., in: Ciba Foundation Symposium on Bone Structure and Metabolism, p. 175. Wolstenholme, G. W. E., O'Conner, C. M., (eds.). Boston: Little, Brown & Co. 1956
45. Rasmussen, H., DeLuca, H., Arnaud, C., Hawker, C., von Stedingk, M.: J. Clin. Invest. *42,* 1940 (1963)
46. Harrison, H. E., Harrison, H. C.: Am. J. Physiol. *201,* 1007 (1961)
47. Morii, H., Lund, J., Neville, P. F., DeLuca, H. F.: Arch. Biochem. Biophys. *120,* 508 (1967)
48. Blunt, J. W., DeLuca, H. F., Schnoes, H. K.: Biochemistry *7,* 3317 (1968)
49. Blunt, J. W., DeLuca, H. F.: Biochemistry *8,* 671 (1969)
50. Suda, T., DeLuca, H. F., Hallick, R. B.: Annal. Biochem. *43,* 139 (1971)
51. DeLuca, H. F., in: The fat-soluble vitamins, p. 3. DeLuca, H. F., Suttie, J. W., (eds.). Madison: University of Wisconsin Press 1970
52. Cousins, R. J., DeLuca, H. F., Suda, T., Chen, T., Tanaka, Y.: Biochemistry *9,* 1453 (1970)
53. Holick, M. F., Schnoes, H. K., DeLuca, H. F.: Proc. Nat. Acad. Sci. USA *68,* 803 (1971)
54. Holick, M. F., Schnoes, H. K., DeLuca, H. F., Gray, R. W., Boyle, I. T., Suda, T.: Biochemistry *11,* 4251 (1972)
55. Semmler, E. J., Holick, M. F., Schnoes, H. K., DeLuca, H. F.: Tetrahedron Letters *40,* 4147 (1972)
56. Paaren, H. E., Schnoes, H. K., DeLuca, H. F.: Chem. Commun. 890 (1977)
57. Ponchon, G., DeLuca, H. F.: J. Clin. Invest. *48,* 1273 (1969)
58. Ponchon, G., Kennan, A. L., DeLuca, H. F.: J. Clin. Invest. *48,* 2032 (1969)
59. Tucker, G., III, Gagnon, R. E., Haussler, M. R.: Arch. Biochem. Biophys. *155,* 47 (1973)
60. Bhattacharyya, M. H., DeLuca, H. F.: Biochem. Biophys. Res. Commun. *59,* 734 (1974)
61. Holick, S. A., Holick, M. F., Tavela, T. E., Schnoes, H. K., DeLuca, H. F.: J. Biol. Chem. *251,* 1025 (1976)
62. Haussler, M. R., Myrtle, J. F., Norman, A. W.: J. Biol. Chem. *243,* 4055 (1968)
63. Lawson, D. E. M., Wilson, P. W., Kodicek, E.: Biochem. J. *115,* 269 (1969)
64. Ponchon, G., DeLuca, H. F.: J. Nutr. *99,* 157 (1969)
65. Haussler, M. R., Boyce, D. W., Littledike, E. T., Rasmussen, H.: Proc. Nat. Acad. Sci. USA *68,* 177 (1971)
66. Fraser, D. R., Kodicek, E.: Nature *228,* 764 (1970)
67. Lawson, D. E. M., Fraser, D. R., Kodicek, E., Morris, H. R., Williams, D. H.: Nature *230,* 228 (1971)
68. Gray, R., Boyle, I., DeLuca, H. F.: Science *172,* 1232 (1971)

69. Norman, A. W., Midgett, R. J., Myrtle, J. F., Nowicki, H. G.: Biochem. Biophys. Res. Commun. *42,* 1082 (1971)
70. Boyle, I. T., Miravet, L., Gray, R. W., Holick, M. F., DeLuca, H. F.: Endocrinology *90,* 605 (1972)
71. Wong, R. G., Norman, A. W., Reddy, C. R., Coburn, J. W.: J. Clin. Invest. *51,* 1287 (1972)
72. Holick, M. F., Garabedian, M., DeLuca, H. F.: Science *176,* 1146 (1972)
73. Chen, T. C., Castillo, L., Korycka-Dahl, M., DeLuca, H. F.: J. Nutr. *104,* 1056 (1974)
74. Garabedian, M., Pavlovitch, H., Fellot, C., Balsan, S.: Proc. Nat. Acad. Sci. USA *71,* 554 (1974)
75. Pechet, M. M., Hesse, R. H.: Am. J. Med. *57,* 13 (1974)
76. Trummel, C. L., Raisz, L. G., Blunt, J. W., DeLuca, H. F.: Science *163,* 1450 (1969)
77. Reynolds, J. J., Holick, M. F., DeLuca, H. F.: Calc. Tiss. Res. *12,* 295 (1973)
78. Olson, E. B., DeLuca, H. F.: Science *165,* 405 (1969)
79. Corradino, R. A.: Science *179,* 402 (1973)
80. Parkes, C. O., Reynolds, J. J.: Mol. Cell. Endocrinol. *7,* 25 (1977)
81. Kream, B. E., Jose, M. J. L., DeLuca, H. F.: Arch, Biochem. Biophys. *179,* 402 (1977)
82. Eisman, J. A., DeLuca, H. F.: Steroids *30,* 245 (1977)
83. Holick, M. F., Garabedian, M., DeLuca, H. F.: Science *176,* 1247 (1972)
84. Holick, M. F., Garabedian, M., DeLuca, H. F.: Biochemistry *11,* 2715 (1972)
85. Holick, M. F., DeLuca, H. F., Kasten, P. M., Korycka, M. B.: Science *180,* 964 (1973)
86. Frolik, C. A., DeLuca, H. F.: J. Clin. Invest. *51,* 2900 (1972)
87. Frolik, C. A., DeLuca, H. F.: Arch. Biochem. Biophys. *147,* 143 (1971)
88. Tsai, H. C., Wong, R. G., Norman, A. W.: J. Biol. Chem. *247,* 5511 (1972)
89. Harnden, D., Kumar, R., Holick, M. F., DeLuca, H. F.: Science *193,* 493 (1976)
90. Kumar, R., Harnden, D., DeLuca, H. F.: Biochemistry *15,* 2420 (1976)
91. Kumar, R., DeLuca, H. F.: Biochem. Biophys. Res. Commun. *69,* 197 (1976)
92. Suda, T., DeLuca, H. F., Schnoes, H. K., Ponchon, G., Tanaka, Y., DeLuca, H. F.: Biochemistry *9,* 2917 (1970)
93. Lam, H.-Y., Schnoes, H. K., DeLuca, H. F., Chen, T. C.: Biochemistry *12,* 4851 (1973)
94. Seki, M., Koizumi, N., Morisaki, M., Ikekawa, N.: Tetrahedron Letters *1,* 15 (1975)
95. Uskokovic, M. R., Baggiolini, E., Mahgoub, A., Narwid, T., Partridge, J. J., in: Vitamin D and problems related to uremic bone disease, p. 279. Norman, A. W., Schaefer, K., Grigoleit, H. G., von Herrath, D., Ritz, E., (eds.). Berlin: Walter de Gruyter 1975
96. Tanaka, Y., DeLuca, H. F., Ikekawa, N., Morisaki, M., Koizumi, N.: Arch. Biochem. Biophys. *170,* 620 (1975)
97. Holick, M. F., Baxter, L. A., Schraufrogel, P. K., Tavela, T. E., DeLuca, H. F.: J. Biol. Chem. *251,* 397 (1976)
98. Boris, A., Hurley, J. F., Trmal, T.: J. Nutr. *107,* 194 (1977)
99. Henry, H. L., Norman, A. W., Taylor, A. N., Hartenbower, D. L., Coburn, J. W.: J. Nutr. *106,* 724 (1976)
100. Friedlander, E. J., Norman, A. W.: Arch. Biochem. Biophys. *170,* 731 (1975)
101. Boyle, I. T., Omdahl, J. L., Gray, R. W., DeLuca, H. F.: J. Biol. Chem. *248,* 4174 (1973)
102. Holick, M. F., Kleiner-Bossaller, A., Schnoes, H. K., Kasten, P. M., Boyle, I. T., DeLuca, H. F.: J. Biol. Chem. *248,* 6691 (1973)
103. Tanaka, Y., Castillo, L., DeLuca, H. F., Ikekawa, N.: J. Biol. Chem. *252,* 1421 (1977)
104. Castillo, L., Tanaka, Y., DeLuca, H. F., Ikekawa, N.: Mineral Electrolyte Metab. *1,* 198 (1978)
105. Bordier, P., Ryckwaert, A., Marie, P., Miravet, L., Norman, A., Rasmussen, H., in: Vitamin D: Biochemical, chemical and clinical aspects related to calcium metabolism, p. 897. Norman, A. W., Schaefer, K., Coburn, J. W., DeLuca, H. F., Fraser, D., Grigoleit, H. G., v. Herrath, D., (eds.). Berlin: Walter de Gruyter, Inc. 1977
106. Henry, H. L., Taylor, A. N., Norman, A. W.: J. Nutr. *107,* 1918 (1977)
107. Canterbury, J. M., Lerman, S., Clafin, A. J., Henry, H., Norman, A., Reiss, E.: J. Clin. Invest. *61,* 1375 (1978)
108. Garabedian, M., Corvol, M. T., Baily du Bois, M., Lieberherr, M., Balsan, S.: Abstract in Proceedings of the Sixth Parathyroid Conference, p. 160. B. C., Vancouver 1977

109. Henry, H. L., Norman, A. W.: Science *201,* 835 (1978)
110. Kanis, J. A., Heynen, G., Russell, R. G. G., Smith, R., Walton, R. J., Warner, G. T., in: Vitamin D: Biochemical, chemical and clinical aspects related to calcium metabolism, p. 794. Norman, A. W., Schaefer, K., Coburn, J. W., DeLuca, H. F., Fraser, D., Grigoleit, H. G., v. Herrath, D., (eds.). Berlin: Walter de Gruyter, Inc. 1977
111. Ikekawa, N., Morisaki, N., Koizumi, N., Sawamura, M., Tanaka, Y., DeLuca, H. F.: Biochem. Biophys. Res. Commun. *62,* 485 (1975)
112. Tanaka, Y., Frank, H., DeLuca, H. F., Koizumi, N., Ikekawa, N.: Biochemistry *14,* 3293 (1975)
113. Tanaka, Y., DeLuca, H. F., Akaiwa, A., Morisaki, M., Ikekawa, N.: Arch. Biochem. Biophys. *177,* 615 (1976)
114. Jones, G., Baxter, L. A., DeLuca, H. F., Schnoes, H. K.: Biochemistry *15,* 713 (1976)
115. Imrie, M. H., Neville, P. F., Snellgrove, A. W., DeLuca, H. F.: Arch. Biochem. Biophys. *120,* 525 (1967)
116. Suda, T., DeLuca, H. F., Schnoes, H. K., Tanaka, Y., Holick, M. F.: Biochemistry *9,* 4776 (1970)
117. Lam, H.Y., Schnoes, H. K., DeLuca, H. F.: Steroids *25,* 247 (1975)
118. Redel, J., Bell, P. A., Bazely, N., Calando, Y., Delbarre, F., Kodicek, E.: Steroids *24,* 463 (1974)
119. Redel, J., Bazely, N., Tanaka, Y., DeLuca, H. F.: FEBS Letters, in press (1978)
120 Suda, T., DeLuca, H. F., Schnoes, H. K., Blunt, J. W.: Biochemistry *8,* 3515 (1969)
121. Jones, G., Schnoes, H. K., DeLuca, H. F.: Biochemistry *14,* 1250 (1975)
122. Hunt, R. D., Garcia, F. G., Hegsted, D. M.: Lab. Animal Care *17,* 222 (1967)
123. Jones, G., Schnoes, H. K., DeLuca, H. F.: J. Biol. Chem. *251,* 24 (1976)
124. Drescher, D., DeLuca, H. F., Imrie, M. H.: Arch. Biochem. Biophys. *130,* 657 (1969)
125. Jones, P. G., Hughes, M. R., Haussler, M. R., in: Vitamin D: Biochemical, chemical and clinical aspects related to calcium metabolism, p. 261. Norman, A. W., Schaefer, K., Coburn, J. W., DeLuca, H. F., Fraser, D., Grigoleit, H. G., v. Herrath, D., (eds.). Berlin: Walter de Gruyter, Inc. 1977
126. Higaki, M., Takahashi, M., Suzuki, T., Sahashi, Y.: J. Vitaminol. *11,* 266 (1965)
127. Higaki, M., Takahashi, M., Suzuki, T., Sahashi, Y.: J. Vitaminol. *11,* 261 (1965)
128. Avioli, L. V., Lee, S. W., McDonald, J. E., Lund, J., DeLuca, H. F.: J. Clin. Invest. *46,* 983 (1967)
129. Kodicek, E., in: Ciba Foundation Symposium on Bone Structure and Metabolism, p. 161. Wolstenholme, G. W. E., O'Connor, C. M., (eds.). Boston: Little, Brown, and Co. 1956
130. Bell, P. A., Kodicek, E.: Biochem. J. *115,* 663 (1969)
131. Avioli, L. V., Williams, T. F., Lund, J., DeLuca, H. F.: J. Clin. Invest. *46,* 1907 (1967)
132. Sahashi, Y., Suzuki, T., Higaki, M., Asano, T.: J. Vitaminol. *13,* 33 (1967)
133. Lakdawala, D., Widdowson, E. M.: The Lancet *I,* 167 (1977)
134. leBoulch, N., Gulat-Marnay, C., Raoul, Y.: Internat. J. Vit. Nutr. Res. *44,* 167 (1974)
135. Sherman, H. C., in: Food products. New York: The Macmillan Company 1933
136. Blondin, G. A., Kulkarni, B. D., Nes, W. R.: Comp. Biochem. Physiol. *20,* 379 (1967)
137. Velluz, L., Amiard, G., Petit, A.: Bull. Soc. Chim. France *16,* 501 (1949)
138. Daniels, F., Jr., in: Handbook of Physiology, p. 969. Baltimore: Williams and Wilkins 1964
139. Esvelt, R. P., Schnoes, H. K., DeLuca, H. F.: Arch. Biochem. Biophys. *188,* 282 (1978)
140. Holick, M. F., Frommer, J. E., McNeill, S. C., Richtand, N. M., Henley, J. W., Potts, J. T. Jr.: Biochem. Biophys. Res. Commun. *76,* 107 (1977)
141. Horsting, M., DeLuca, H. F.: Biochem. Biophys. Res. Commun. *36,* 251 (1969)
142. Olson, E. B. Jr., Knutson, J. C., Bhattacharyya, M. H., DeLuca, H. F.: J. Clin. Invest. *57,* 1213 (1976)
143. Jones, G.: Clin. Chem. *24,* 287 (1978)
144. Bhattacharyya, M., DeLuca, H. F.: Arch. Biochem. Biophys. *160,* 58 (1974)
145. Ritter, M. C., Dempsey, M. E.: J. Biol. Chem. *246,* 1536 (1971)
146. Scallen, T. J., Schuster, M. W., Dhat, A. K.: J. Biol. Chem. *246,* 224 (1971)
147. Björkhem, I., Holmberg, I.: J. Biol. Chem. *253,* 842 (1978)
148. Bhattacharyya, M. H., DeLuca, H. F.: J. Biol. Chem. *248,* 2969 (1973)

149. Ponchon, G., DeLuca, H. F., Suda, T.: Arch. Biochem. Biophys. *141,* 397 (1970)
150. Norman, A. W.: Am. J. Med. *57,* 21 (1974)
151. Haddad, J. G., Stamp, T. C. B.: Am. J. Med. *57,* 57 (1974)
152. Belsey, R., Clar, M. B., Bernat, M., Glowacki, J., Holick, M. F., DeLuca, H. F., Potts, J. T. Jr.: Am. J. Med. *57,* 50 (1974)
153. Madhok, T. C., Schnoes, H. K., DeLuca, H. F.: Biochem. J. *175,* 479 (1978)
154. Schaefer, K., Flury, W. H., Von Herrath, D., Kraft, D., Sweingruber, R.: Schweiz. Med. Wochschr. *102,* 785 (1972)
155. Silver, J., Neal, G., Thompson, G. R.: Clin. Sci, Mol. Med. *46,* 443 (1974)
156. Hahn, T. J., Hendin, B. A., Scharp, C. R., Haddad, J. G.,Jr.: New Engl. J. Med. *287,* 900 (1972)
157. Lifshitz, F., McLaren, N.: J. Pediat. *83,* 612 (1973)
158. Hahn, T. J., Birge, S. J., Scharp, C. R., Avioli, L. V.: J. Clin. Invest. *51,* 741 (1972)
159. Sulimovici, S., Roginsky, M. S.: Life Sci. *21,* 1317 (1977)
160. Gray, R. W., Omdahl, J. L., Ghazarian, J. G., DeLuca, H. F.: J. Biol. Chem. *247,* 7528 (1972)
161. Ghazarian, J. G., Schnoes, H. K., DeLuca, H. F.: Biochemistry *12,* 2555 (1973)
162. Ghazarian, J. G., DeLuca, H. F.: Arch. Biochem. Biophys. *160,* 63 (1974)
163. Henry, H. L., Norman, A. W.: J. Biol. Chem. *249,* 7529 (1974)
164. Ghazarian, J. G., Jefcoate, C. R.: Knutson, J. C., Orme-Johnson, DeLuca, H. F.: J. Biol. Chem. *249,* 3026 (1974)
165. Pedersen, J. I., Ghazarian, J. G., Orme-Johnson, N. R., DeLuca, H. F.: J. Biol. Chem. *251,* 3933 (1976)
166. Knutson, J. C., DeLuca, H. F.: Biochemistry *13,* 1543 (1974)
167. Kumar, R., Schnoes, H. K., DeLuca, H. F.: J. Biol. Chem. *253,* 3804 (1978)
168. Garabedian, M., Corvol, M.-T., Nguyen, T. M., Balsan, S.: Ann. Biol. Anim. Biochim. Biophys. *18,* 175 (1978)
169. Madhok, T. C., Schnoes, H. K., DeLuca, H. F.: Biochemistry *16,* 2142 (1977)
170. Tanaka, Y., DeLuca, H. F.: Science *183,* 1198 (1974)
171. Tanaka, Y., Lorenc, R. S., DeLuca, H. F.: Arch. Biochem. Biophys. *171,* 521 (1975)
172. Tanaka, Y., Shepard, R. M., DeLuca, H. F., Schnoes, H. K.: Biochem. Biophys. Res. Commun. *83,* 7 (1978)
173. Boyle, I. T., Gray, R. W., Omdahl, J. L., DeLuca, H. F., in: Endocrinology 1971, p. 468. Taylor, S. (ed.). London: Wm. Heinemann Medical Books, Ltd. 1972
174. Boyle, I. T., Gray, R. W., DeLuca, H. F.: Proc. Nat. Acad. Sci. *68,* 2131 (1971)
175. Tanaka, Y., DeLuca, H. F.: Arch. Biochem. Biophys. *154,* 566 (1973)
176. Omdahl, J. L., Gray, R. W., Boyle, I. T., Knutson, J., DeLuca, H. F.: Nature (New Biology) *237,* 63 (1972)
177. Henry, H. L., Midgett, R. J., Norman, A. W.: J. Biol. Chem. *249,* 7584 (1974)
178. Garabedian, M., Holick, M. F., DeLuca, H. F., Boyle, I. T.: Proc. Nat. Acad. Sci. *69,* 1673 (1972)
179. Fraser, D. R., Kodicek, E.: Nature (New Biology) *241,* 163 (1973)
180. Booth, B. E., Tsai, H. C., Morris, C.: J. Clin. Invest. *60,* 1314 (1977)
181. Suda, T., Horiuchi, N., Fukushima, M., Nishii, Y., Ogata, E., in: Vitamin D: Biochemical, chemical and clinical aspects related to calcium metabolism, p. 201. Norman, A. W., Schaefer, K., Coburn, J. W., DeLuca, H. F., Fraser, D., Grigoleit, H. G., von Herrath, D., (eds.). Berlin: Walter de Gruyter, Inc. 1977
182. Rasmussen, H., Wong, M., Bikle, D., Goodman, D. B. P.: J. Clin. Invest. *51,* 2502 (1972)
183. Shain, S. A.: J. Biol. Chem. *247,* 4404 (1972)
184. Larkins, R. G., MacAuley, S. J., Rapoport, A., Martin, T. J., Tulloch, B. R., Byfield, P. G. H., Matthews, E. W., MacIntyre, I.: Clin. Sci. Mol. Med. *46,* 569 (1974)
185. Galante, L., MacAuley, S. J., Colston, K. W., MacIntyre, I.: Lancet *I,* 985 (1972)
186. Galante, L., Colston, K. W., MacAuley, S. J., MacIntyre, I.: Nature *238,* 271 (1972)
187. Lorenc, R., Tanaka, Y., DeLuca, H. F., Jones, G.: Endocrinol. *100,* 468 (1977)
188. Colston, K. W., Evans, I. M. A., Galante, L., MacIntyre, I., Moss, D. W.: Biochem. J. *134,* 817 (1973)

189. Bikle, D. D., Rasmussen, H.: J. Clin. Invest. *55,* 292 (1975)
190. Bikle, D. D., Murphy, E. W., Rasmussen, H.: J. Clin. Invest. *55,* 299 (1975)
191. Suda, T., Horiuchi, N., Sasaki, S., Ogata, E., Ezawa, I., Nagata, N., Kimura, S.: Biochem. Biophys. Res. Commun. *54,* 512 (1973)
192. Tanaka, Y., Chen. T. C., DeLuca, H. F.: Arch. Biochem. Biophys. *152,* 291 (1972)
193. Tsai, H. C., Midgett, R. J., Norman, A. W.: Arch. Biochem. Biophys. *157,* 339 (1973)
194. Juan, D., DeLuca, H. F.: Endocrinology *101,* 1184 (1977)
195. Larkins, R. G., MacAuley, S. J., MacIntyre, I.: Nature *252,* 412 (1974)
196. Colston, K. W., Evans, I. M. A., Spelsberg, T. C., MacIntyre, I.: Biochem. J. *164,* 83 (1977)
197. Larkins, R. G., MacAuley, S. J., MacIntyre, I.: Mol. Cell. Endocrinol. *2,* 193 (1975)
198. Chen, T. C., DeLuca, H. F.: Arch. Biochem. Biophys. *156,* 321 (1973)
199. Baxter, L. A., DeLuca, H. F.: J. Biol. Chem. *251,* 3158 (1976)
200. Ribovich, M. L., DeLuca, H. F.: Arch. Biochem. Biophys. *170,* 529 (1975)
201. Bar, A., Wasserman, R. H.: Biochem. Biophys. Res. Commun. *54,* 191 (1973)
202. Steele, T. H., DeLuca, H. F.: J. Clin. Invest. *57,* 867 (1976)
203. DeLuca, H. F.: Acta Orthop. Scand. *46,* 286 (1975)
204. Potts, J. T., Jr., in: Parathyroid hormone and thyrocalcitonin (calcitonin), p. 403. Talmage, R. V., Belanger, L. F. (eds.). Amsterdam: Excerpta Medica 1968
205. Zull, J. E., Repke, D. W.: J. Biol. Chem. *247,* 2915 (1972)
206. Garabedian, M., Tanaka, Y., Holick, M. F., DeLuca, H. F.: Endocrinology *94,* 1022 (1974)
207. Arnaud, C. D.: Am. J. Med. *55,* 577 (1973)
208. Corradino, R. A., Ebel, J. G., Craig, P. H., Taylor, A. N., Wasserman, R. H.: Calc. Tiss. Res. *7,* 81 (1971)
209. Omdahl, J. L., DeLuca, H. F.: Science *174,* 949 (1971)
210. Omdahl, J. L., DeLuca, H. F.: J. Biol. Chem. *247,* 5520 (1972)
211. Omdahl, J. L., Hunsaker, L. A., Aschenbrenner, V. A.: Arch. Biochem. Biophys. *184,* 172 (1977)
212. Bonjour, J.-P., Russell, R. G. G., Morgan, D. B., Fleisch, H. A.: Europ. J. Clin. Invest. *2,* 274 (1972)
213. Baxter, L. A., DeLuca, H. F., Bonjour, J.-P., Fleisch, H.: Arch. Biochem. Biophys. *164,* 655 (1974)
214. Hill, L. F., Lumb, G. A., Mawer, E. B., Stanbury, S. W.: Clin. Sci. *44,* 335 (1973)
215. Bonjour, J.-P., Trechsel, U., Fleisch, H., Schenk, R., DeLuca, H. F., Baxter, L. A.: Am. J. Physiol. *229,* 402 (1975)
216. Baxter, L. A.: Ph. D. Thesis, University of Wisconsin (1976)
217. Botham, K. M., Tanaka, Y., DeLuca, H. F.: Biochemistry *13,* 4961 (1974)
218. Botham, K. M., Ghazarian, J. G., Kream, B. E., DeLuca, H. F.: Biochemistry *15,* 2130 (1976)
219. Ghazarian, J. G., Kream, B. E., Botham, K. M., Nickells, M. W., DeLuca, H. F.: Arch. Biochem. Biophys. *189,* 212 (1978)
220. Kream, B. E., DeLuca, H. F., Moriarity, D. M., Kendrick, N. C., Ghazarian, J. G.: Arch. Biochem. Biophys. *192,* 318 (1979)
221. Boass, A., Toverud, S. U., McCain, T. A., Pike, J. W., Haussler, M. R.: Nature *267,* 630 (1977)
222. Haussler, M. R., McCain, T. A.: New England J. Med. *297,* 974 (1977)
223. Spanos, E., MacIntyre, I.: Lancet *1,* 840 (1977)
224. Riddle, O., Reinhart, W. H.: Am. J. Physiol. *76,* 660 (1926)
225. Common, R. H.: J. Agric. Sci. *23,* 555 (1933)
226. Kyes, P., Potter, T. S.: Anat. Rec. *60,* 377 (1934)
227. Common, R. H., Rutledge, N. A., Hale, R. W.: J. Agric. Sci. *38,* 64 (1948)
228. Tanaka, Y., Castillo, L., DeLuca, H. F.: Proc. Nat. Acad. Sci. *73,* 2701 (1976)
229. Castillo, L., Tanaka, Y., DeLuca, H. F., Sunde, M. L.: Arch. Biochem. Biophys. *179,* 211 (1977)
230. Kenny, A. D.: Am. J. Physiol. *230,* 1609 (1976)
231. Tanaka, Y., Castillo, L., Wineland, M. J., DeLuca, H. F.: Endrocrinology *103,* 2035 (1978)
232. Baksi, S. N., Kenny, A. D.: Biochem. Pharmacol. *26,* 2439 (1977)

233. Baksi, S. N., Kenny, A. D.: Endrocrinology *101,* 1216 (1977)
234. Avioli, L. V., Birge, S. J., Lee, S. W.: J. Clin. Endocrinol. *28,* 1341 (1968)
235. Favus, M. J., Walling, M. W., Kimberg, D. V.: J. Clin. Invest. *52,* 1680 (1973)
236. Carre, M., Ayigbede, O., Miravet, L., Rasmussen, H.: Proc. Nat. Acad. Sci. *71,* 2996 (1974)
237. Holick, M. F., DeLuca, H. F., in: Advances in steroid biochemistry and pharmacology, pp. 111–155. Briggs, M. H., Christie, G. A. (eds.). London and New York: Academic Press 1974
238. Eisman, J. A., Shepard, R. M., DeLuca, H. F.: Anal. Biochem. *80,* 298 (1977)
239. Haddad, J. G., Jr., Min, C., Mendelsohn, M., Slatopolsky, E., Hahn, T. J.: Arch. Biochem. Biophys. *182,* 390 (1977)
240. Horst, R. L., Shepard, R. M., Jorgensen, N. A., DeLuca, H. F.: Arch. Biochem. Biophys. *192,* 512 (1979)
241. Haddad, J. G., Chyu, K. J.: J. Clin. Endocr. *33,* 992 (1971)
242. Belsey, R., DeLuca, H. F., Potts, J. T., Jr.: J. Clin. Endocrinol. Metab. *33,* 554 (1971)
243. Horst, R. L., Shepard, R. M., Jorgensen, N. A., DeLuca, H. F.: J. Lab. Clin. Med. *93,* 277 (1979)
244. Eisman, J. A., Hamstra, A. J., Kream, B. E., DeLuca, H. F.: Arch. Biochem. Biophys. *176,* 235 (1976)
245. Brumbaugh, P. F., Haussler, D. H., Bressler, R., Haussler, M. R.: Science *183,* 1089 (1974)
246. Stamp, T. C. B., Round, J. M., Rowe, D. J. F., Haddad, J. G.: Brit. Med. J. *4,* 9 (1972)
247. DeLuca, H. F., Avioli, L. V., in: Renal disease. Black, D., (ed.). Oxford: Blackwell Scientific Publications, Ltd., 1978
248. Fraser, D., Kooh, S. W., Kind, H. P., Holick, M. F., Tanaka, Y., DeLuca, H. F.: New England J. Med. *289,* 817 (1973)
249. DeLuca, H. F.: Arch. Internat'l. Med. *138,* 836 (1978)
250. DeLuca, H. F.: Proc. Ann. Royal Coll. Physicians & Surgeons January, pp. 216–225 (1977)
251. Coburn, J. W., Hartenbower, D. L., Brickman, A. S.: Am. J. Clin. Nutr. *29,* 1283 (1976)
252. Hahn, T. J., Avioli, L. V., in: Vitamin D: Biochemical, chemical, and clinical aspects related to calcium metabolism, pp. 737–745. Norman, A. W., Schaefer, K., Coburn, J. W., DeLuca, H. F., Fraser, D., Grigoleit, H. G., von Herrath, D., (eds.). Berlin: Walter de Gruyter, Inc. 1977
253. Wasserman, R. H., Kallfelz, F. A., Comar, C. L.: Science *133,* 883 (1961)
254. Schachter, D., in: The transfer of calcium and strontium across biological membranes, pp. 197–210. Wasserman, R. H. (ed.). New York: Academic Press 1963
255. Martin, D. L., DeLuca, H. F.: Arch. Biochem. Biophys. *134,* 139 (1969)
256. Walling, M. H., Rothman, S. S.: Am. J. Physiol. *217,* 1144 (1969)
257. Schachter, D., Rosen, S. M.: Am. J. Physiol. *196,* 357 (1959)
258. Harrison, H. C., Harrison, H. E.: Am. J. Physiol. *217,* 121 (1969)
259. Martin, D. L., DeLuca, H. F.: Am. J. Physiol. *216,* 1351 (1969)
260. Schachter, D., Kowarski, S., Finkelstein, J. D., Wang Ma, R.: Am. J. Physiol. *211,* 1131 (1966)
261. Borle, A. B.: Endocrinology *88A,* 155 (1971)
262. Zull, J. E., Czarnowska-Misztal, E., DeLuca, H. F.: Science *149,* 182 (1965)
263. Tanaka, Y., DeLuca, H. F., Omdahl, J., Holick, M. F.: Proc. Nat. Acad. Sci. *68,* 1286 (1971)
264. Bikle, D. D., Zolock, D. T., Morrissey, R. L., Herman, R. H.: J. Biol. Chem. *253,* 484 (1978)
265. Tanaka, Y., DeLuca, H. F.: Proc. Nat. Acad. Sci. *68,* 605 (1971)
266. Wasserman, R. H., Feher, J. J., in: Calcium-binding proteins and calcium function, pp. 293–303. Wasserman, R. H., Corradino, R. A., Carafoli, E., Kretsinger, R. H., MacLennan, D. H., Siegel, F. L., (eds.). North Holland: Elsevier Press 1977
267. Morrissey, R. L., Wasserman, R. H.: Am. J. Physiol. *220,* 1509 (1971)
268. Wasserman, R. H., Corradino, R. A.: Ann. Rev. Biochem. *40,* 501 (1971)
269. Eilon, G., Mor, E., Karaman, H., Menczel, J., in: Cellular mechanisms for calcium transfer and homeostasis, pp. 501–502. New York: Academic Press 1971
270. Harmeyer, J., DeLuca, H. F.: Arch. Biochem. Biophys. *133,* 247 (1969)
271. Spencer, R., Charman, M., Wilson, P. W., Lawson, D. E. M.: Biochem. J. *170,* 93 (1978)
272. Spencer, R., Charman, M., Wilson, P. W., Lawson, D. E. M.: Nature *263,* 161 (1976)

273. Wasserman, R. H., Taylor, A. N., Fulmer, C. S.: Biochem. Soc. Spec. Publ. *3*, 55 (1974)
274. Corradino, R. A.: J. Cell Biol. *58*, 64 (1973)
275. Corradino, R. A.: Nature *243*, 41 (1973)
276. Moriuchi, S., DeLuca, H. F.: Arch. Biochem. Biophys. *164*, 165 (1974)
277. Taylor, A. N., Wasserman, R. H.: J. Histochem. Cytochem. *18*, 107 (1970)
278. Morrissey, R. L., Empson, R. N., Jr., Zolock, D. T., Bikle, D. D., Bucci, T. J.: Biochim. Biophys. Acta *538*, 34 (1978)
279. Chen, T. C., Weber, J. C., DeLuca, H. F.: J. Biol. Chem. *245*, 3776 (1970)
280. Haussler, M. R., Norman, A. W.: Proc. Nat. Acad. Sci. *62*, 155 (1969)
281. Lawson, D. E. M., Wilson, P. W., Barker, D. C., Kodicek, E.: Biochem. J. *115*, 263 (1969)
282. Chen, T. C., DeLuca, H. F.: J. Biol. Chem. *248*, 4890 (1973)
283. Yamada, S., Schnoes, H. K., DeLuca, H. F.: Analyt. Biochem. *85*, 34 (1978)
284. Zile, M., Bunge, E. C., Barsness, L., Yamada, S., Schnoes, H. K., DeLuca, H. F.: Arch. Biochem. Biophys. *186*, 15 (1978)
285. Haddad, J. G., Jr., Birge, S. J.: Biochem. Biophys. Res. Commun. *45*, 829 (1971)
286. Haddad, J. G., Hahn, T. J., Birge, S. J.: Biochim. Biophys. Acta *329*, 93 (1973)
287. Haddad, J. G., Birge, S. J.: J. Biol. Chem. *250*, 299 (1975)
288. Kream, B. E., Reynolds, R. D., Knutson, J. C., Eisman, J. A., DeLuca, H. F.: Arch. Biochem. Biophys. *176*, 779 (1976)
289. Van Baelen, H., Bouillon, R., DeMoor, P.: J. Biol. Chem. *252*, 2515 (1977)
290. Hughes, M. R., Haussler, M. R.: J. Biol. Chem. *253*, 1065 (1978)
291. Chertow, B. S., Baylink, D. J., Wergedal, J. E., Su, M. H. H., Norman, A. W.: J. Clin. Invest. *56*, 668 (1975)
292. Tanaka, Y., DeLuca, H. F., Ghazarian, J. G., Hargis, G. K., Williams, G. A.: Mineral & Electrolyte Metab. (in press) (1978)
293. Brumbaugh, P. F., Haussler, M. R.: Life Sciences *13*, 1737 (1973)
294. Brumbaugh, P. F., Haussler, M. R.: Life Sciences *16*, 353 (1975)
295. Brumbaugh, P. F., Haussler, M. R.: J. Biol. Chem. *250*, 1588 (1975)
296. Brumbaugh, P. F., Haussler, M. R.: J. Biol. Chem. *249*, 1258 (1974)
297. Brumbaugh, P. F., Haussler, M. R.: J. Biol. Chem. *249*, 1251 (1974)
298. Kream, B. E., DeLuca, H. F.: Biochem. Biophys. Res. Commun. *76*, 735 (1977)
299. Kream, B. E., Yamada, S., Schnoes, H. K., DeLuca, H. F.: J. Biol. Chem. *252*, 4501 (1977)
300. Kream, B. E., Jose, M., Yamada, S., DeLuca, H. F.: Science *197*, 1086 (1977)
301. McCain, T. A., Haussler, M. R., Okrent, D., Hughes, M. R.: FEBS Lett. *86*, 65 (1978)
302. Eisman, J. A., Hamstra, A. J., Kream, B. E., DeLuca, H. F.: Science *193*, 1021 (1976)
303. Wilson, P. W., Lawson, D. E. M.: Biochim. Biophys. Acta *497*, 805 (1977)
304. Emtage, J. S., Lawson, D. E. M., Kodicek, E.: Nature *246*, 100 (1973)
305. Max, E. E., Goodman, D. B. P., Rasmussen, H.: Biochim. Biophys. Acta *511*, 224 (1978)
306. Borle, A. B.: J. Membrane Biol. *16*, 207 (1974)
307. Taylor, A. N.: J. Nutr. *104*, 489 (1974)
308. Wasserman, R. H., Taylor, A. N.: J. Nutr. *103*, 586 (1973)
309. Kowarski, S., Schachter, D.: J. Biol. Chem. *244*, 211 (1969)
310. Walling, M. W., in: Vitamin D: Biochemical, chemical and clinical aspects related to calcium metabolism, pp. 321–330. Norman, A. W., Schaefer, K., Coburn, J. W., DeLuca, H. F., Fraser, D., Grigoleit, H. G., von Herrath, D., (eds.). Berlin: Walter de Gruyter, Inc. (1977)
311. DeLuca, H. F.: Vitamins and Hormones *25*, 315 (1967)
312. Tanaka, Y., DeLuca, H. F.: Arch. Biochem. Biophys. *146*, 574 (1971)
313. Frost, H. M., in: Henry Ford Hospital Surgical Monograph Series, Springfield, Charles A. Thomas Col., 1966
314. Raisz, L. G., Trummel, C. L., Holick, M. F., DeLuca, H. F.: Science *175*, 768 (1972)
315. Reynolds, J. J., Holick, M. F., DeLuca, H. F.: Calc. Tiss. Res. *15*, 333 (1974)
316. Stern, P. H., DeLuca, H. F., Ikekawa, N.: Biochem. Biophys. Res. Commun. *67*, 965 (1975)
317. Gran, F. C.: Acta Physiol. Scand. *50*, 132 (1960)
318. Steele, T. H., Engle, J. E., Tanaka, Y., Lorenc, R. S., Dudgeon, K. L., DeLuca, H. F.: Am. J. Physiol. *229*, 489 (1975)
319. Hermsdorf, C. L., Bronner, F.: Biochim. Biophys. Acta *379*, 553 (1975)

320. Taylor, A. N., Wasserman, R. H.: Am. J. Physiol. *223,* 110 (1972)
321. Sutton, R. A. L., Harris, C. A., Wong, N. L. M., Dirks, J., in: Vitamin D: Biochemical, chemical and clinical aspects related to calcium metabolism, pp. 451–453. Norman, A. W., Schaefer, K., Coburn, J. W., DeLuca, H. F., Fraser, D., Grigoleit, H. G., von Herrath, D., (eds.). Berlin: Walter de Gruyter, Inc. 1977
322. Harrison, H. E., Harrison, H. C.: J. Clin. Invest. *20,* 47 (1941)
323. Puschett, J. B., Fernandex, P. C., Boyle, I. T., Gray, R. W., Omdahl, J. L., DeLuca, H. F.: Proc. Soc. Exptl. Biol. Med. *141,* 379 (1972)
324. Popovtzer, M. M., Robinette, J. B., DeLuca, H. F., Holick, M. F.: J. Clin. Invest. *53,* 913 (1974)
325. Brodehl, J., Kaas, W. P., Weber, H.-P.: Pediat. Res. *5,* 591 (1971)
326. Tanaka, Y., DeLuca, H. F.: Proc. Nat. Acad. Sci. *71,* 1040 (1974)
327. Castillo, L., Tanaka, Y., DeLuca, H. F.: Endocrinology *97,* 995 (1975)
328. Campbell, J. A., Squires, D. M., Babcock, H. C.: Steroids *13,* 567 (1969)
329. Halkes, S. J., van Vliet, N. P.: Recl. Trav. Chim. Pays-Bas *88,* 1080 (1969)
330. Redel, J., Bazely, N., Calando, Y., Delbarre, F., Bell, P. A., Kodicek, E.: J. Steroid Biochem. *6,* 117 (1975)
331. Cesario, M., Guilhem, J., Pascard, C., Redel, J.: Tetrahedron Lett. p. 1097 (1978)
332. Morisaki, M., Rubio-Lightbourn, J., Ikekawa, N.: Chem. Pharm. Bull. *21,* 457 (1973)
333. Redel, J., Bell, P., Delbarre, F., Kodicek, E.: C. R. Hebd. Seances Acad. Sci., Ser. D *278,* 529 (1974)
334. Seki, M., Rubio-Lightbourn, J., Morisaki, M., Ikekawa, N.: Chem. Pharm. Bull. *21,* 2783 (1973)
335. Rotman, A., Mazur, Y.: J. Chem. Soc. Chem. Commun. *15,* (1974)
336. Wicha, J., Bal, K.: J. Chem. Soc., Chem. Commun. *968,* (1975)
337. McMorris, T., Schow, S. R.: J. Org. Chem. *41,* 3759 (1976)
338. Narwid, T. A., Cooney, K. E., Uskokovic, M. R.: Helv. Chim. Acta *57,* 771 (1974)
339. Partridge, J. J., Faber, S., Uskokovic, M. R.: Helv. Chim. Acta *57,* 764 (1974)
340. Eyley, S. C., Williams, D. H.: J. Chem. Soc., Perkin Trans. *1,* 727 (1976)
341. Dasgupta, S. K., Gut, M.: J. Org. Chem. *40,* 1475 (1975)
342. Dasgupta, S. K., Crump, D. R., Gut, M.: J. Org. Chem. *39,* 1658 (1974)
343. Partridge, J. J., Toome, V., Uskokovic, M. R.: J. Am. Chem. Soc. *98,* 3739 (1976)
344. Koizumi, N., Morisaki, M., Ikekawa, N.: Tetrahedron Lett. 2899 (1978)
345. Koizumi, N., Morisaka, M., Ikekawa, N., Suzaki, A., Takeshita, T.: Tetrahedron Lett. 2203 (1975)
346. Onisko, B. L., Schnoes, H. K., DeLuca, H. F.: Tetrahedron Lett. 1107 (1977)
347. Narwid, T. A., Blount, J. F., Iacobelli, J. A., Uskokovic, M. R.: Helv. Chim. Acta *47,* 781 (1974)
348. Barton, D. H. R., Hesse, R. H., Pechet, M. M., Rizzardo, E.: J. Am. Chem. Soc. *95,* 2748 (1973)
349. Barton, D. H. R., Hesse, R. H., Pechet, M. M., Rizzardo, E.: J. Chem. Soc. Chem. Commun. 203 (1974)
350. Ikekawa, N., Morisaki, M., Koizumi, N., Kato, Y., Takeshita, T.: Chem. Pharm. Bull. *23,* 695 (1975)
351. Lam, H.-Y., Schnoes, H. K., DeLuca, H. F.: Science *186,* 1038 (1974)
352. Morisaki, M., Koizumi, N., Ikekawa, N., Takeshita, T., Ishimoto, S.: J. Chem. Soc. Perkin Trans. *1,* 1421 (1975)
353. Lam, H.-Y., Schnoes, H. K., DeLuca, H. F., Reeve, L., Stern, P. H.: Steroids *26,* 422 (1975)
354. Sheves, M., Berman, E., Freeman, D., Mazur, Y.: J. Chem. Soc., Chem. Commun. 643 (1975)
355. Okamura, W., Pirio, M. R.: Tetrahedron Lett. 4317 (1975)
356. Kaneko, C., Yamada, S., Sugimoto, A., Eguchi, Y., Ishikawa, M., Suda, T., Suzuki, M., Kakuta, S., Sasaki, S.: Steroids *23,* 75 (1973)
357. Kaneko, C., Yamada, S., Sugimoto, A., Ishikawa, M., Sasaki, S., Suda, T.: Tetrahedron Lett. 2339 (1973)

358. Kaneko, C., Sugimoto, A., Eguchi, Y., Yamada, S., Ishikawa, M., Sasaki, S., Suda, T.: Tetrahedron *30,* 2701 (1974)
359. Mihailović, M., Lorenc, L., Pavlović, V., Kalvoda, J.: Tetrahedron *33,* 441 (1977)
360. Aberhart, D. J., Chu, J. Y.-R., Hsu, A. C.-T.: J. Org. Chem. *41,* 1067 (1976)
361. Sheves, M., Mazur, Y.: Tetrahedron Lett. 1913 (1976)
362. Okamura, W. H., Mitra, M. N., Wing, R. M., Norman, A. W.: Biochem. Biophys. Res. Commun. *60,* 179 (1974)
363. Onisko, B. L., Lam, H. Y., Reeve, L., Schnoes, H. K., DeLuca, H. F.: Bioorg. Chem. *6,* 203 (1977)
364. Hammond, M. L.. Mouriño, A., Okamura, W. H.: J. Am. Chem. Soc. *100,* 4907 (1978)
365. Bakker, S. A., Lugtenburg, J., Havinga, E.: Recl. Trav. Chim. Pays-Bas *91,* 1459 (1972)
366. Inhoffen, H. H., Kath, J., Sticherling, W., Brücher, K.: Ann. Chemie *603,* 25 (1957)
367. Harrison, R. G., Lythgoe, B., Wright, P. W.: J. Chem. Soc., Perkin Trans. *1,* 2654 (1974)
368. Westerhof, P., Keverling-Buisman, J. A.: Recl. Trav. Chim. Pays-Bas *75,* 453 (1956)
369. Mourino, A., Okamura, W. H.: J. Org. Chem. *43,* 1653 (1978)
370. Paaren, H. E.: Ph. D. Thesis, University of Illinois, Chicago Circle (1976)
371. Napoli, J. L., Fivizzani, M. A., Schnoes, H. K., DeLuca, H. F.: Biochemistry *17,* 2387 (1978)
372. Napoli, J. L., Fivizzani, M. A., Schnoes, H. K., DeLuca, H. F.: Biochemistry (in press) (1979)
373. Sheves, M., Sialom, B., Mazur, Y.: J. Chem. Soc., Chem. Commun. 554 (1978)
374. Loibner, H., Zbiral, E.: Tetrahedron *34,* 713 (1978)
375. Sheves, M., Friedmann, N., Mazur, Y.: J. Org. Chem. *43,* 3597 (1977)
376. Pelc, B.: Steroids *30,* 193 (1977)
377. Paaren, H. E., Schnoes, H. K., DeLuca, H. F.: Proc. Nat. Acad. Sci. *75,* 2080 (1978)
378. Hallick, R. B., DeLuca, H. F.: J. Biol. Chem. *246,* 5733 (1971)
379. Muccino, R. R., Vernici, G. G., Cupano, J., Oliveto, E. P., Liebman, A. A.: Steroids *31,* 645 (1978)
380. Sheves, M., Mazur, Y.: J. Am. Chem. Soc. *97,* 6249 (1975)
381. Kaneko, C., Yamada, S., Sugimoto, A., Ishikawa, M., Suda, T., Suzuki, M., Sasaki, S.: J. Chem. Soc., Perkin Trans. *1,* 1104 (1975)
382. Bosman, H. B., Chen, P. S., Jr.: J. Nutr. *90,* 141 (1966)
383. Holick, M. F., Garabedian, M., Schnoes, H. K., DeLuca, H. F.: J. Biol. Chem. *250,* 226 (1975)
384. Windaus, A., Trautmann, G.: Hoppe-Seylers Z. Physiol. Chem. *247,* 185 (1937)
385. DeLuca, H. F., Weller, M., Blunt, J. W., Neville, P. F.: Arch. Biochem. Biophys. *124,* 122 (1968)
386. Crump, D. R., Williams, D. H., Pelc, B.: J. Chem. Soc., Perkin Trans. *1,* 2731 (1973)
387. Bontekoe, J. S., Wignall, A., Rappoldt, M. P., Roborgh, J. R.: Int. Z. Vitaminforsch. *40,* 589 (1970)
388. Johnson, R. L., Carey, S. C., Norman, A. W., Okamura, W. H.: J. Med. Chem. *20,* 5 (1977)
389. Mourino, A., Blair, P., Wecksler, W., Johnson, R. L., Norman, A., W., Okamura, W. H.: J. Med. Chem. *21,* 1025 (1978)
390. Fürst, A., Labler, L., Meier, W., Pfoertner, K. H.: Helv. Chim. Acta *56,* 1108 (1973)
391. Holick, M. F., Semmler, E. J., Schnoes, H. K., DeLuca, H. F.: Science *180,* 190 (1973)
392. Okamura, W. H., Mitra, M. N., Procsal, D. A., Norman, A. W.: Biochem. Biophys. Res. Commun. *65,* 24 (1975)
393. Schnoes, H. K., DeLuca, H. F.: Vitam. Horm. (N. Y.) *32,* 385 (1974)
394. Berman, E., Friedman, N., Mazur, Y., Sheves, M.: J. Am. Chem. Soc. *100,* 5626 (1978)
395. Zerwekh, J. E., Brumbaugh, P. F., Haussler, D. H., Cork, D. J., Haussler, M. R.: Biochemistry *13,* 4097 (1974)
396. Procsal, D. A., Okamura, W. H., Norman, A. W.: J. Biol. Chem. *250,* 8382 (1975)
397. Okamura, W. H., Mitra, M. N., Procsal, D. A., Norman, A. W.: Biochem. Biophys. Res. Commun. *65,* 24 (1975)
398. Stern, P. H., Trummel, C. L., Schnoes, H. K., DeLuca, H. F.: Endocrinology *97,* 1552 (1975)

399. Stern, P. H., Mavreas, T., Trummel, C. L., Schnoes, H. K., DeLuca, H. F.: Mol. Pharmacol. *12,* 879 (1976)
400. Mahgoub, Sheppard, H.: Endocrinology *100,* 629 (1977)
401. Lam, P. H.-Y., Reeve, L., Mellon, W. S., Stern, P. H., DeLuca, H. F.: Steroids, (in press) (1979)
402. Trummel, C. L., Raisz, L. G., Hallick, R. B., DeLuca, H. F.: Biochem. Biophys. Res. Commun. *44,* 1096 (1971)
403. DeLuca, H. F., Schnoes, H. K.: Ann. Rev. Biochem. *46,* 631 (1976)
404. Schnoes, H. K., DeLuca, H. F., in: Bioorganic chemistry, p. 299. van Tamalen, E. E., (ed.). New York: Academic Press 1978
405. Tanaka, Y., Frank, H., DeLuca, H. F.: Endocrinology *92,* 417 (1973)
406. Haussler, M. R., Zerwekh, J. E., Hesse, R. H., Rizzardo, E., Pechet, M. M.: Proc. Nat. Acad. Sci. *70,* 2248 (1973)
407. Holick, M. F., Kasten-Schraufrogel, P., Tavela, T., DeLuca, H. F.: Arch. Biochem. Biophys. *166,* 63 (1976)
408. Fukishima, K., Suzuki, Y., Tohira, Y., Matsunaga, I., Ochi, K., Nagano, H., Nishii, Y., Suda, T.: Biochem. Biophys. Res. Commun. *66,* 632 (1975)
409. Holick, M. F., Tavela, T., Holick, S. A., Schnoes, H. K., DeLuca, H. F., Gallagher, B. M.: J. Biol. Chem. *251,* 1020 (1971)
410. Hallick, R. B., DeLuca, H. F.: J. Biol. Chem. *246,* 5733 (1971)
411. Lawson, D. E. M., Bell, T. A.: Biochem. J. *142,* 37 (1974)
412. Napoli, J. L., Mellon, W. S., Fivizzani, M. A., Schnoes, H. K., DeLuca, H. F.: J. Biol. Chem. (in press) 1979
413. Lawson, D. E. M., Friedman, N., Sheves, M., Mazur, Y.: FEBS Letters *80,* 137 (1977)

Received November 27, 1978

Cytochrome P450 and Biological Hydroxylation Reactions

Volker Ullrich

Department of Physiological Chemistry, University of Saarland, 6650 Homburg-Saar, Germany

Table of Contents

1 Introduction

The essential role of molecular oxygen, now termed dioxygen, as an energy providing source in aerobic living systems has always been evident but the mechanism by which this energy is used to drive biological oxidations has been controversial for many years. In the mid 1930's O. Warburg[1] concluded from model studies of iron-catalyzed oxidations with molecular oxygen that an activation of the dioxygen molecule was the crucial event for biological oxidations, which at that time were still rather ill defined and thought mainly to consist of glycolysis and mitochondrial respiration. In contrast, H. Wieland[2] was the proponent of the dehydrogenation hypothesis which considered the oxygen molecule merely as an electron sink for substrate hydrogen. It was this mechanism which proved to be correct for all energy producing reactions in the cell. Warburg's idea of oxygen activation was almost forgotten until in 1955 Mason[3] and Hayaishi[4] demonstrated, using the stable isotope $^{18}O_2$, that biological oxidations which proceed by a direct incorporation of dioxygen into organic compounds are rather common in nature. These findings initiated a boom in the research on the corresponding enzymic reactions. It is now established that these oxygenations are catalyzed by a variety of dioxygen activating enzymes, which were designated "oxygenases" by Hayaishi[5]. Depending on whether both oxygen atoms or only one atom are introduced into the substrate, a subclassification into "dioxygenases" and "monooxygenases" was proposed and the following stoichiometry established:

$$RH + O_2 \longrightarrow RO_2H \text{ (diooxygenases)} \quad (1)$$
$$RH + O_2 + DH_2 \longrightarrow ROH + D + H_2O \text{ (monooxygenases)} \quad (2)$$

The latter enzymes had earlier been called "mixed function oxidases" by Mason[6] in order to characterize their double functions as oxygenases and oxidases. For the reduction of one oxygen atom of molecular oxygen to water two electrons from an external donor have to be supplied. Sometimes the substrate itself can provide these reducing equivalents, so that for these enzymes the term "internal monooxygenases" may be used[7].

Whereas dioxygenases usually require substrates activated by functional groups, the monooxygenases can act upon chemically rather inert compounds like steroids or aromatic rings and convert them to the corresponding alcohols or phenols. In fact, this is the only mechanism existing in nature to convert aliphatic or aromatic compounds to other functional derivatives. Therefore, these enzymes play an important role in metabolism and many of them have been described in recent years. To appreciate fully the catalytic power of monooxygenases one has to consider the thermodynamics of the monooxygenation process as calculated for a secondary CH-bond as in cyclohexane[8]:

$$NADPH + H^+ \longrightarrow H_2 + NADP^+ \qquad +\ 19.3\ \text{kJ/mol} \quad (3)$$
$$^3O_2\ (g) \longrightarrow 2\ O\ (g) \qquad +460.5\ \text{kJ/mol} \quad (4)$$
$$H_2\ (g) + O\ (g) \longrightarrow H_2O\ (fl) \qquad -470.1\ \text{kJ/mol} \quad (5)$$
$$AH\ (fl) + O\ (g) \longrightarrow AOH\ (fl) \qquad -393.9\ \text{kJ/mol} \quad (6)$$
$$AH + O_2 + NADPH + H^+ \longrightarrow AOH + H_2O + NADP^+ \qquad -384.2\ \text{kJ/mol} \quad (7)$$

Two extremely high energy barriers have to be overcome for the reaction: firstly, the dissociation of the O–O bond (457 kJ/mol) and secondly the dissociation of the CH bond, in the order of 418 kJ/mol. This exemplifies the importance of enzyme catalysis in general, and explains why chemists have been fascinated by the ease, specificity and economy of the hydroxylation process in particular. Many efforts have been undertaken to copy the monooxygenase mechanism in chemical models, but the results have not been very satisfactory (see Chap. 9). It soon became evident that first a complete knowledge of the enzymes was required before a corresponding chemical oxygenase reaction could be set up. Unfortunately, the isolation of monooxygenase systems proved to be complicated because of their multicomponent nature and membrane-bound structure. An important step forward was the recognition that many monooxygenase systems contained the same type of biocatalyst, which was characterized as a hemoprotein called "cytochrome P450".

2 Occurrence and Function of Cytochrome P450

As in many other cases in modern science the discovery of cytochrome P450 and its biological role followed a sequence of apparently not directly related findings. It started in 1958 with the description by Garfinkel[9] and Klingenberg[10] of a pigment in the microsomal fraction liver which was characterized by an unusual carbon monoxide absorption difference spectrum with a peak at 450 nm. This finding was made possible by the sophisticated spectrophotometric techniques for turbid solutions developed by B. Chance at the Johnson Research Foundation in Philadelphia. A typical difference spectrum of reduced microsomes with carbon monoxide is shown in Fig. 1.

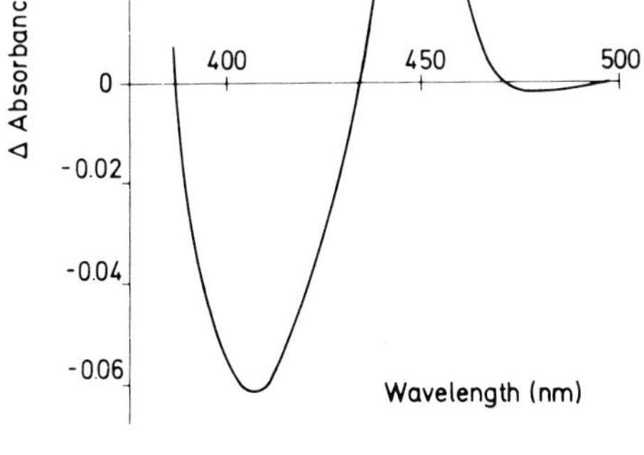

Fig. 1. Difference spectrum of reduced liver microsomes with carbon monoxide

At that time no correlation was suggested for this pigment with the capacity of the endoplasmic reticulum of liver to convert drugs and other foreign compounds to hydroxylated derivatives in the presence of dioxygen and NADPH[11]. A possible relation of the 450 nm absorbing carbon monoxide complex with enzymatic hydroxylation reactions became very likely, when Ryan and Engel[12] observed the carbon monoxide inhibition of the C-21 steroid hydroxylase in bovine adrenal microsomes. Proof for this assumption was presented by Estabrook, Cooper and Rosenthal[13], who elegantly used the photochemical action spectrum of the CO-inhibited C-21-hydroxylase to show the involvement of the pigment in the oxygen activation process of the hydroxylase system. Shortly after, Omura and Sato[14, 15] characterized the pigment as a hemoprotein with protoporphyrin IX as the prosthetic group. However, attempts to solubilize the hemoprotein resulted in a loss of the typical absorption of the CO-complex at 450 nm with a concomitant shift to 420 nm. The CO-complexes of most hemoproteins absorb at this wavelength, so that an unusual coordination sphere of the native hemoprotein was suggested which evidently was lost upon denaturation. The native protein was termed "cytochrome P450" by Omura and Sato, whereas the inactive form was referred to as "cytochrome P420".

In the following years cytochrome P450 was also found as an integral part of the 11β-hydroxylase system in adrenal mitochondria[16]. Its involvement in the liver microsomal monooxygenation of codeine, acetanilide[17] and cyclohexane[18] was established using the technique of the photochemical action spectrum (Fig. 2).

Using the same technique many more monooxygenase systems have been reported to contain cytochrome P450 as the oxygen activating component. Its occurrence is not restricted to mammalian tissues; fish[19], birds[20], yeast[21, 22], plants[23] and bacteria[24] also contain cytochrome P450-dependent monooxygenases. Although not all of the enzymes have been fully characterized, a generalization of the functions

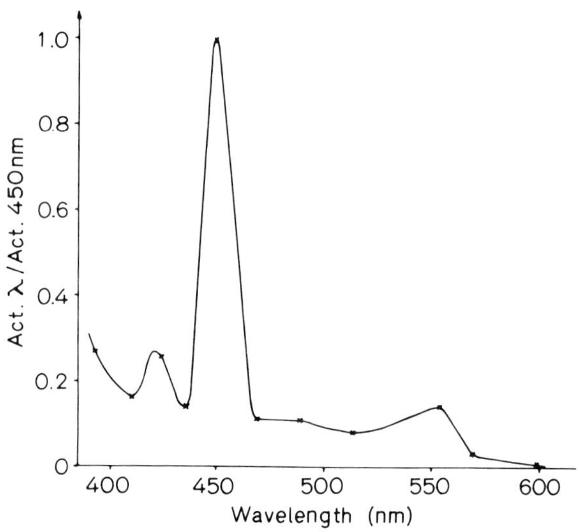

Fig. 2. Photochemical action spectrum of the reversibility of the carbon monoxide inhibited cyclohexane hydroxylation in rat liver microsomes. From[18]

Table 1. Cytochrome P450-dependent monooxygenases

Organism/organ	Subcellular localization	Substrate	Product	Ref.
Bacteria				
Ps. putida	Cytosol	Camphor	5-OH-exo-camphor	27)
Bacillus megaterium	Cytosol	3-Oxo-Δ^4steroid	15β-OH-steroids	28)
Fungi				
Cunninghamella bainieri	Microsomes	Xenobiotics	Alcohol, phenols epoxides	29)
Rhizopus nigricans	Microsomes	Progesterone	11α-OH-progesterone	30)
Plants				
Vinca rosea	Microsomes	Geraniol	10-OH-geraniol	31)
Echinocystis macrocarpa green endosperm	Microsomes	Kaur-16-ene	Kaurenol	23)
Yeasts				
S. cerevisiae	Microsomes	Xenobiotics	Alcohols, phenols, epoxides	32)
Candida tropicalis	Microsomes	Alkanes, fatty acids	n-Alcohols ω-OH-fatty acids	21, 22)
Insects				
Musca domestica	Microsomes	Xenobiotics	Alcohols, phenols epoxides	33)
Manduca sexta	Mitochondria	α-Ecdysone	β-Ecdysone	34)
Mammals				
Adrenal gland	Microsomes	Steroids	21-OH-steroids	13)
Adrenal gland	Mitochondria	Deoxycorticosterone	Cortisol	35)
Liver	Microsomes	Xenobiotics	Alcohols, phenols epoxides	36, 37)
Liver	Mitochondria	Bile acids precursors	26-OH bile acids	38)
Liver	Microsomes	Cholesterol	7α-OH-precursors	39)
Kidney	Microsomes	25-OH-vitamin D_3	1,25-di-OH-vitamin D_3	40)
Kidney cortex	Microsomes	Fatty acids	ω-OH fatty acids	41)

of cytochrome P450 seems possible. So far, it is known to be involved in cholesterol and steroid metabolism in different cells and organisms[25] and in the monooxygenation of drugs and "xenobiotics", as all compounds foreign to living organisms have been named by Mason et al.[26]. No function in amino acid metabolism has been discovered up to now. Table 1 contains a list of cytochrome P450-dependent monooxygenases with their origin, subcellular localizations and functions.

Not only does the table show that cytochrome P450-dependent monooxygenases are present in all types of organisms but, excepting bacteria, they are localized either on the inner mitochondrial or on the endoplasmic reticulum membranes. As will be indicated in Chap. 5 this association with the two different membranes is also connected with two different pathways of reduction of cytochrome P450. The reduction in mitochondria involves a flavoprotein and an iron-sulfur protein, whereas in microsomal systems the reduction occurs directly by a flavoprotein of a more complex structure. It is of interest to note that the reducing system in bacteria corresponds to that of mitochondria, which leads support to the bacterial origin of these organelles.

Another characteristic feature of cytochrome P450-dependent monooxygenases can also be seen on Table 1: the variation in specificities towards substrates and the variety of products. Some monooxygenases tolerate only small modifications in the shape of the substrate molecule and the introduction of the oxygen atom with these enzymes occurs with high stereospecificity. Others seem to exhibit a rather broad specificity and their products are multiple and point to an only limited stereoselectivity. The reason for this is not entirely known, but a teleological explanation seems to be that enzymes with endogenous substrates and important functions in the biosynthesis of steroids or hormones must be more specific since a well-defined product is required for further functions in the organism. In contrast those enzymes involved in the degradation of xenobiotics have to metabolize a broad spectrum of lipophilic, foreign compounds, for which the introduction of an oxygen atom at any position in the molecules is usually sufficient to increase the hydrophilic properties and conjugation ability of the compound and hence to facilitate its elimination from the body.

A special situation applies to microorganisms such as yeasts and bacteria, which use organic chemicals from the environment as the sole carbon source for growth. If a microorganism grows on a pure hydrocarbon a monooxygenation is the only mechanism that can initiate its degradation. The corresponding monooxygenases must therefore exhibit high specificity to ensure a specific metabolic pathway for further degradation and a high molecular activity to ensure sufficient growth rates. The most wellknown system in this respect is the camphor-5-exo-monooxygenase from Ps. putida, which uses only a small number of structurally closely related camphor analogues as substrates and specifically introduces the hydroxyl group in the 5-exo-position[42]. Its turnover rate is around 1000/min and therefore is one magnitude higher than those of the drug monooxygenases in mammalian liver[43]. The bacterial enzyme also has a high affinity binding for substrate[44] suggesting a specific and stereoselective fit of the camphor molecule to the active site of the protein.

In the case of the nonspecific drug monooxygenase system in liver and other tissues of many organisms one would expect a rather loose binding of the various drug substrates if a common binding site at only one enzyme were to exist. This, however, would be at variance with the required ability of the enzyme to remove even small concentrations of toxic chemicals from the body. The answer to this dilemma was found after the isolation of monooxygenase systems from livers and lungs of various species. This showed that several cytochrome P450 forms exist with slightly different, but overlapping substrate specifities. For example one form can exhibit a rather high selectivity for a certain group of substrates but not for others. The relative distribution of the different cytochrome P450 forms depends on genetic and environmental influences (see Chap. 9). The numerous observations of differences in drug metabolism when parameters like sex, age, disease, strain, species, nutritional state or exposure to chemicals are considered find a reasonable explanation in the multiple forms of cytochrome P450 enzymes. This implies that the substrate binding site is an essential and integral part of cytochrome P450. The biochemical investigation of the structure of cytochrome P450 has lent further support to this assumption.

3 Structure of Cytochrome P450

If the various forms of cytochrome P450 have different specificities for substrates, they must also differ in amino acid composition. It is now established that in all forms so far characterised, differences in molecular weight, amino acid composition and terminal amino acids exist. Figure 3 shows the optical absorption spectra of cytochrome P450 from Ps. putida.

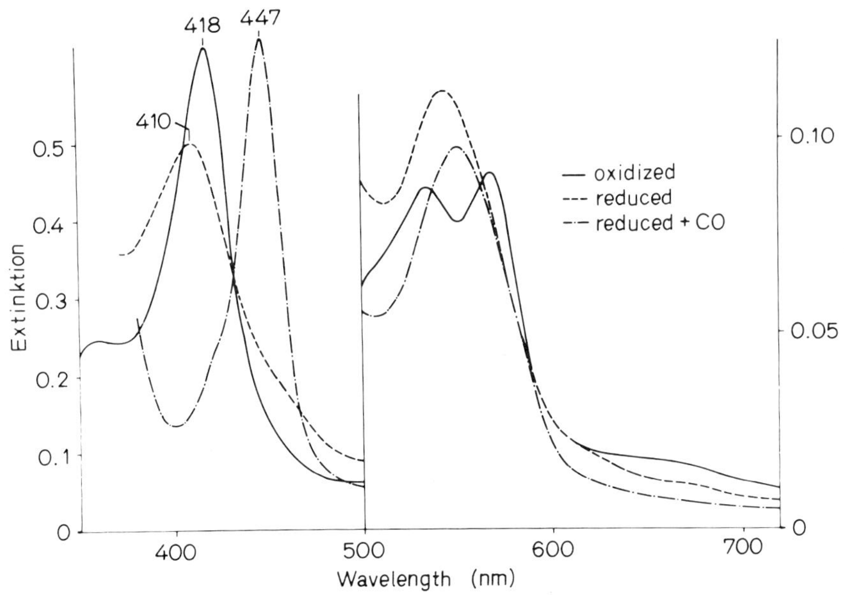

Fig. 3. Optical spectra of cytochrome P450 from camphor grown Ps. putida. The sample was a courtesy of Dr. J. A. Peterson

Compared to other hemoproteins the spectra are unusual with respect to the blue-shift and absorption decrease upon reduction. Addition of carbon monoxide to the reduced hemoprotein results in formation of the 447 nm Soret absorption band which appears at 450 nm in a difference spectrum and gave its name to this cytochrome.

Antibodies prepared against pure cytochrome P450 proteins in general show no cross-reactivities[45)]. On the other hand the spectral properties are almost identical and only very small, but significant, differences between the various forms have been reported[46)]. The heme is bound in all cytochrome P450 forms to a single polypeptide chain with a molecular weight between 40000 and 60000 D. The similarity in the optical spectra could be interpreted as an essentially identical coordination sphere of the heme, which is only slightly influenced by the residual amino acid composition. From the photochemical action spectrum it is evident that the heme in its unusual environment is the dioxygen activating site and hence the catalytic center of the enzyme. Since the loss of the characteristic optical absorption spectra is paralleled by a loss in activity, it may be deduced that the clue to the unique oxygen activation process lies in the special coordination sphere of cytochrome P450.

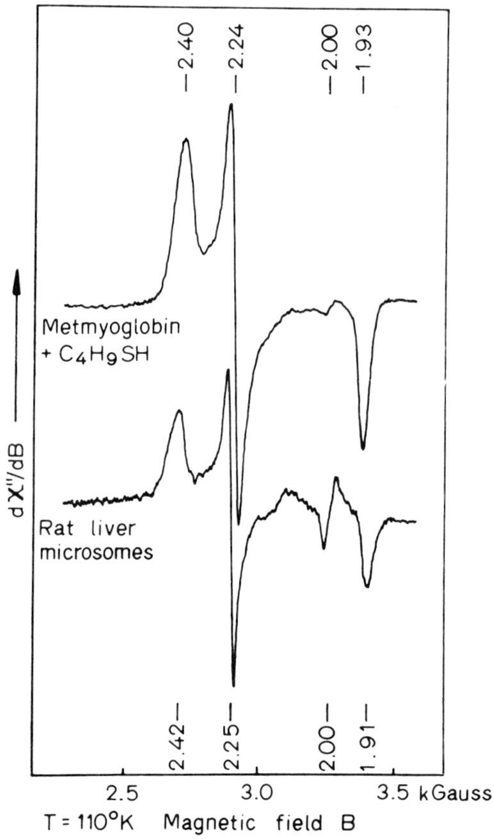

Fig. 4. EPR spectra of cytochrome P450 in rat liver microsomes (bottom) and a metmyoglobin-n-butane thiol complex[47)]

Out of the six ligands in the octahedral field of the heme iron, four are pyrrole nitrogens of protoporphyrin IX. Therefore, the unusual coordination sphere must be determined by the fifth and/or sixth ligands, which in other hemoproteins are usually donated by the protein. The first indication of the nature of one of the protein ligands came from EPR spectra of modified metmyoglobin complexes. All forms of cytochrome P450 show very similar EPR spectra with g-values at liquid nitrogen temperature around g = 1.92, 2.24 and 2.44. This rather narrow splitting can be simulated when metmyoglobin is incubated with an alkane thiol, such as butane thiol[47] (Fig. 4).

Synthetic heme complexes with mercaptides as one ligand also show very similar EPR characteristics, indicating that a mercaptide group from a cysteine sulfur would most likely be the fifth ligand in cytochrome P450[49–51]. This assumption is further substantiated when organic ligand molecules are added to cytochrome P450 in its oxidized state. Whereas the EPR-spectra show only small variations in g-values with the various ligands, the optical absorption changes profoundly especially in the Soret region (Table 2).

Table 2. Spectroscopic parameters of cytochrome $P450_{CAM}$-ligand complexes

Ligand		g-values by EPR			Bands in optical spectra (nm)
		g_x	g_y	g_z	
Native (?)	Ox.	2.45	2.26	1.91	417 535 571
Native	Red.		–		411 540
Carbon monoxide	Red.		–		447 550
Octylamine	Ox. Red.	2.494	2.255 –	1.894	423 538 574 445
Metyrapone[a)]	Ox. Red.	2.48	2.26 –	1.88	421 536 442 539 566
1-Methyl-imidazole	Ox. Red.	2.535	2.259 –	1.873	424 540 578
Ethyl isocyanide	Ox. Red.	2.49	2.32 –	1.89	430 548 453 549 576
Octylmethylsulfide	Ox. Red.	2.46	2.26 –	1.90	426 537 572 447 542
Diethylphenylphosphine	Ox. Red.	2.50	2.28 –	1.88	377 454 556 460
Benzylthiolate	Ox. Red.	2.37 (43)	2.25 –	1.94 (2)	377 465 557 –

a) 2-Methyl-1,2-di-3-pyridyl-1-propanone.

As can be seen from Table 2 when phosphines or thiols were used as ligands with ferric cytochrome, the Soret absorption splits and shifts towards the far red region of the spectrum. In the reduced state all ligands except oxygen lead to a Soret absorption band around 450 nm similar to that of the reduced CO-complex. Absorption measurements with polarized light of the crystallized CO-complex indicated that the axial ligand was responsible for the unusual spectra[52)]. In porphyrin chemistry similar spectra exist and are classified as "hyperporphyrin" spectra[53)]. They are supposed to originate from a charge transfer from an electronegative ligand to the central atom. This again supports the suggestion of a mercaptide in cytochrome P450. Model complexes consisting of iron-porphyrin thiolates are difficult to obtain, since at ambient temperature the ferric complexes are readily reduced and in the ferrous state their affinity for mercaptides seems to be rather low[54)]. Stern and Peisach[55)] succeeded in preparing a heme carbon monoxide complex with a mercaptide ligand by adding a large excess of the thiol in very alkaline solution. At lower temperatures of about 200 °K the reduction of ferric porphyrin complexes is slowed down considerably, allowing the preparation of five- and six-coordinated ferric thiolate complexes of lipid-soluble porphyrins. At the temperature of liquid nitrogen these complexes are stable and their optical and EPR-spectra have been studied (Table 3).

It is evident from a comparison with the data of Table 2 that the anomalous spectra of ferric cytochrome P450 could be mimicked almost exactly under these conditions, leaving no doubt about the thiolate nature of its fifth ligand. Additional proof comes from Mößbauer studies, which show an unusual temperature dependence of the quadrupole splitting[56)], which was also found in the model complexes with a thiolate ligand. Furthermore, the model complexes listed in Table 3 can be used as the basis for speculation on the nature of the sixth ligand in low-spin ferric cytochrome P450.

Its Soret absorption at 416–418 nm rules out soft ligands and indicates that a nitrogen ligand, as postulated by Chevion et a.[57)] on the basis of a comparison of ligand parameters of cytochrome P450 with model complexes, is less likely since

Table 3. Spectroscopic parameters of hemin-toluene thiolate ligand complexes (Ruf, Sakurai, Ullrich, unpublished). To the five coordinated heme toluene thiolate complex the other ligands were added

6th Ligand	g-Values by EPR			Bands in optical spectra (nm)
	g_x	g_y	g_z	
Toluene thiolate	2.302	2.229	1.959	376 470 561
Diethylphenylphosphine	2.391	2.264	1.924	374 458 556
Pentamethylenesulfide	2.333	2.248	1.946	(365) 432 533 578
Benzylisocyanide	2.322	2.246	1.942	432 536 558
n-Octylamine	2.391	2.239	1.931	429 539 571
Imidazole	2.363	2.241	1.937	428 538 568
Pyridine	2.362	2.253	1.935	424 536 564
Methanol	2.411	2.273	1.926	417 534 566

the Soret bands of such complexes always absorb above 420 nm. They would also be tightly binding ligands in the reduced state showing 450 nm bands, which is in contrast to the observed weak Soret absorption band of reduced cytochrome P450 at around 414 nm. NMR-pulse relaxation studies have indicated the presence of a proton at a distance of 2.8–3.0 Å from the iron, which also would exclude histidine as the sixth ligand[58]. From the model studies it is more likely that an OH-group of either a serine, threonine or water occupies this position since the EPR and especially the optical spectra with OH ligands are similar to low spin ferric cytochrome P450[59]. Since hydroxyl oxygen only binds with the ferric and not with the ferrous cytochrome, it would not compete with dioxygen binding the reduced state. This would be in agreement with the concept, that in the presence of substrates the sixth ligand is removed from the heme leading to a pentacoordinated hemoprotein. This species is high-spin and has a 30 nm blue-shifted Soret absorption and a band at 645 nm. The interconversion of the low spin complex to the high-spin form by a substrate is shown in Fig. 5 for the cytochrome $P450_{CAM}$.

A model complex with properties almost identical to the high spin complex was obtained in a crystalline form from hemin-dimethylester and 4-nitrobenzene thiolate[61]. By X-ray analysis it proved to be a five-coordinated complex with an Fe-S distance of 2.32 Å[61]. A low affinity sixth ligand would thus allow a low-spin/high-spin equilibrium with an equilibrium constant close to unity. Both substrates and an increase in temperature would shift this equilibrium to the high-spin side. The significance of this mechanism for enzymatic catalysis will be discussed in Chap. 6.1.

Compared to an Fe-N linkage to the protein, an Fe-S bond is more labile, which might explain the instability of cytochrome P450 enzymes to denaturating conditions. The charge transfer from the mercaptide, being responsible for the splitting of the Soret band, would be easily lost upon changing the distance between the iron and sulfur atom[62]. This probably occurs in the presence of detergents and leads to a loss in the 450 nm-absorption of the reduced CO-complex with concomitant formation of a 420 nm absorption band. Surprisingly, the characteristic g-values in the

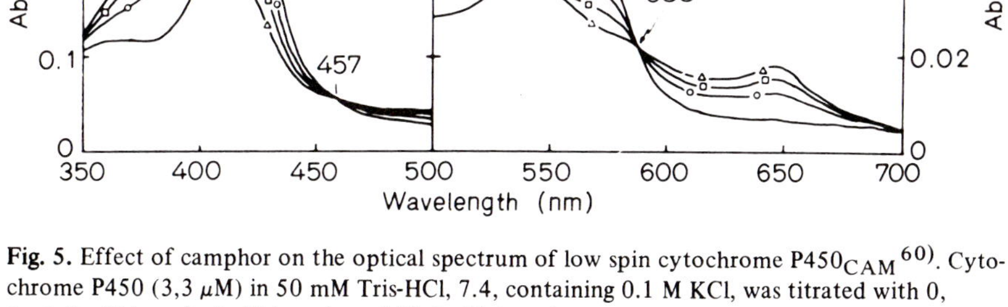

Fig. 5. Effect of camphor on the optical spectrum of low spin cytochrome $P450_{CAM}$[60]. Cytochrome P450 (3,3 μM) in 50 mM Tris-HCl, 7.4, containing 0.1 M KCl, was titrated with 0, (—); 2 μM (–o–), 6 μM (–□–), and 20.5 μM (–Δ–)

EPR-spectrum of the oxidized cytochrome are maintained, which may indicate that the stronger interaction of the ferric complex is more resistant to destruction than the reduced form with its weaker Fe^{2+}-S^{-} bonding. This state of inactivation seems to be prevented in the presence of glycerol, so that purification of cytochrome P450-depentent monooxygenases have been successful when working in 20–30% glycerol solutions. More drastic denaturating conditions, like acid treatment or mercurials, lead to an inactivation, which is characterized by a disappearance of the typical low spin signal of cytochrome P450 under formation of a high spin g = 6 signal[63].

It should be noted that another hemoprotein, chloroperoxidase, has spectral characteristics very similar to cytochrome P450[64]. This enzyme activates chloride ions by the use of hydrogen peroxide and has no monooxygenase functions. Surprisingly, no cysteine can be detected after amino acid analysis, so that the role of mercaptide sulfur in producing the characteristic hyperporphyrin spectra has been queried[65]. However, the enzyme is known to contain a cysteine disulfide bridge and it has not yet been excluded whether this has been formed by a heme catalyzed oxidation of two cysteines during the course of the experiments. Thus the possibility remains open that chloroperoxidase is indeed a heme-sulfur protein and that the mercaptide group plays a role in the chloride activation process.

4 Nomenclature

The existence of multiple cytochrome P450 enzymes has revealed that the nomenclature of these monooxygenases is no longer consistent with the rules of the Enzyme Commission. First, the essential features of cytochrome P450 are substrate binding and dioxygen activation rather than electron transfer as defined for cytochromes. Therefore, these proteins should be considered more as enzymes than as electron-transfer proteins. The wavelength in subscript ("450") should indicate the position of the α-band of the heme and not that of the Soret band. In addition the currently used characterization of a form of cytochrome P450 by the known differences in the CO-absorption band (e. g. P448, P449, P452) has been found unsatisfactory since more forms absorb at a given wavelength. Another reason to abandon the term cytochrome is the new definition of cytochromes, which requires strong field ligands in the fifth and sixth coordination position of the hemochrome[35]. This probably does not apply for the sixth ligand position of cytochrome P450.

When named as enzymes the various cytochromes P450 should be termed monooxygenases with the name of the substrate and position attached, like steroid-11β-monooxygenase. If the thiolate nature of the protein ligand is unequivocally established the qualification "heme-sulfur-dependent" could be made in parenthesis. Difficulties, however, must arise for the designation of unspecific microsomal monooxygenases, for which many substrates and multiple products can be found. Either these enzymes are termed "xenobiotic monooxygenases" and one includes their source and electrophoretic mobility, or one establishes the best substrate and the position of highest attack. Thus the major form of cytochrome P450 present in rat liver microsomes after pretreatment with methylcholanthrene could be designated as "rat xenobiotic monooxygenase, LM (liver microsomes) 4" (since it is band No. 4 in the gel electropherogram), or "rat liver acetanilide-4-monooxygenase". The disadvantage of the first proposal would be the possible existence of further bands in the electropherogram and in the second case a better substrate could well be found. Since the latter is more likely and subscripts a, b, c etc. could be used for further forms, the first possibility may be of advantage. This would also come closer to the present practice of naming the above mentioned enzyme "rat cytochrome P450 LM4"[67]. In this review the term "cytochrome P450" will be used synonymously with "heme-sulfur-containing" monooxygenases.

5 Electron Donors

An essential step in the mechanism of dioxygen activation is the electron transfer from the reduced pyridine nucleotide to the hemoprotein. So far, two principally different reducing systems have been found. One occurs in bacteria and mitochondria and contains an FAD-flavoprotein together with an iron-sulfur protein. The second consists of an FAD-FMN flavoprotein which directly reduces the hemoprotein without the help of an iron-sulfur protein (Fig. 6).

Both electron transfer chains cannot substitute each other in reconstituted systems, which points to a different phylogenetic origin of mitochondrial and bacterial monooxygenases on the one hand and the microsomal enzymes on the other. Microsomes also contain cytochrome b_5 which is able to donate electrons to the monooxygenases. However, the physiological role of this cytochrome is still controversial and more than one function has to be considered.

5.1 Iron-Sulfur Proteins

The first monooxygenase system isolated was the steroid-11β-monooxygenase from adrenal mitochondria[16]. In a reconstituted system the heme-sulfur protein required an iron-sulfur protein, a flavoprotein and NADPH for reduction and dioxygen acti-

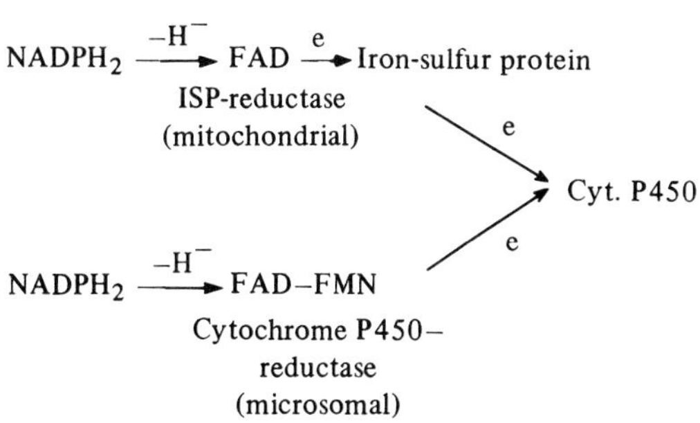

Fig. 6. Mitochondrial and microsomal electron transport for cytochrome P450-dependent monooxygenases

vation. The structure of the iron-sulfur component turned out very similar to that of the known plant ferredoxins and therefore was called " adrenodoxin"[68]. Its molecular weight is 12638 D[69] and it contains a two-iron-two acid-labile sulfur center which accepts and donates only one electron. In the reduced state it exhibits a characteristic g = 1.94 signal in the EPR-spectrum originating from an antiferromagnetic coupling of a high-spin ferric and high-spin ferrous atom[70]. The potential is between −360 and −380 mV and thus is more negative than the standard potential of the NADPH/$NADP^+$ couple[71]. An essentially similar type of iron-sulfur protein was isolated from camphor-grown Ps. putida bacteria, which contain a heme-sulfur dependent camphor-5-exo-monooxygenase[72]. All components are soluble, so that the interaction of the reduced "putidaredoxin" with the hemoprotein could be studied. As a result protein-protein interaction was established by various methods[42, 73]. Reduction of the cytochrome P450 only occurred in the presence of camphor which is most likely to be due to an increase in redox potential after addition of the substrate. Other redoxins were found in mammalian mitochondrial monooxygenase systems with various functions. Liver mitochondria contain an iron-sulfur protein associated with a cholesterol-26-monooxygenase involved in bile acid formation[74]. Similarly, the 1α-hydroxylation of 25-hydroxyvitamin D requires a mitochondrial monooxygenase localized in kidney, from which a renal ferredoxin could be isolated[75]. The reduction of the iron-sulfur proteins by NADPH or NADH is mediated in all systems by FAD-containing reductases having a molecular weight generally around 40000 D in one polypeptide chain. The stoichiometry in which the three components of the monooxygenase system occur in vivo, is close to 1 : 1 : 1, although in reconstituted systems a large excess of the iron-sulfur proteins is required to obtain turnover rates comparable to those observed in the intact organelles.

5.2 Flavoproteins

In contrast to the bacterial and mitochondrial monooxygenase systems, all microsomal enzymes require only one electron transfer component. This protein has been isolated from liver microsomes[76] and only recently was characterized as an FAD and FMN containing flavoprotein[77]. Originally it was termed NADPH-cytochrome c reductase, but later its physiological substrate cytochrome P450 was recognized and therefore this name was changed to cytochrome P450 reductase. It is active towards cytochrome P450, however, only after solubilization with detergents and in the presence of small amounts of phospholipid, preferentially phosphatidyl choline[78]. When the enzyme is cleaved off the membrane by proteolytic procedures, it loses a hydrophobic part of the polypeptide chain, which is needed for the interaction with cytochrome P450, but not with cytochrome c. Similarly, if the FMN is carefully removed from the enzyme it still retains the capacity of reducing cytochrome c but not cytochrome P450[79].

The flavoprotein can be reduced by NADPH to the fully reduced state ($FMNH_2$, $FADH_2$) and is oxidized by ferricyanide to the fully oxidized state (FMN, FAD)[77, 79]. By EPR an O_2-stable semiquinone radical can be observed aerobically in the presence of NADPH which by ferricyanide titration was found to be the one electron reduced form. The higher reduced forms can react with O_2 to give O_2^- radicals, so that titrations of the enzyme with NADPH have to be carried out anaerobically. In the presence of the mediator indigosulfonate the flavins could be titrated by following their optical absorption and the following scheme was derived[77]:

$$F_1F_2 \rightleftharpoons F_1H \cdot F_2 \qquad E_{0,1}' = -0.110\ V \tag{8}$$

$$F_1H \cdot F_2 \rightleftharpoons F_1H_2F_2 \qquad E_{0,2}' = -0.270\ V \tag{9}$$

$$F_1H_2F_2 \rightleftharpoons F_1H_2F_2H \cdot \qquad E_{0,3}' = -0.290\ V \tag{10}$$

$$F_1H_2F_2H \rightleftharpoons F_1H_2F_2H_2 \qquad E_{0,4}' = -0.365\ V \tag{11}$$

Which redox state donates the electrons to cytochrome P450 has not yet been clearly established. According to the redox potential of the cytochrome P450-substrate complex (Chap. 6.1) all states except the one-electron reduced could be involved in the reduction process. Iyanagi and Mason[77] have suggested that the fully reduced state (E′ = –365 mV) may act like an iron-sulfur protein and may be the actual donor for cytochrome P450.

In the reconstituted systems the reductase proved not to be specific for a particular cytochrome P450. The enzyme from liver could reduce the ω-monooxygenase from kidney and *vice versa*[80] and its antibody was effective in inhibiting the NADPH-mediated cytochrome c reduction in spleen microsomes.

5.3 Cytochrome b_5

Liver microsomes in addition to the xenobiotic monooxygenase system contain a second electron transport system. It is specific for NADH and reduces cytochrome b_5 via an FAD-containing flavoprotein (Fig. 7).

The system has a high affinity for NADH[81] and since the redox potential of cytochrome b_5 is about –80 mV[82], it is fully reduced under the conditions of the cell. Using antibodies against NADH-cytochrome b_5 reductase, cytochrome b_5 and NADPH-cytochrome P450 reductase, is was shown that NADH can support the monooxygenation of drugs in rat liver microsomes via cytochrome b_5 and NADH-cytochrome b_5 reductase without the involvement of NADPH-cytochrome P450 reductase[83]. The activity is, however, only 10–30% of that mediated by NADPH. In the presence of both NADH and NADPH a synergistic effect is observed[84] which has been interpreted in two ways: Hildebrandt and Estabrook[85] conclude that the

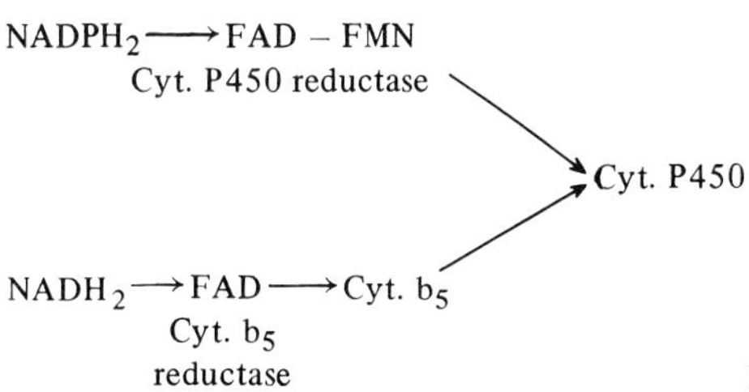

Fig. 7. Microsomal electron transport systems

second electron for dioxygen activation comes from cytochrome b_5 whereas Staudt et al.[86] suggested that the synergistic action of NADH is the result of an electron sparing effect. Two electron consuming reactions may contribute to such an effect:
(i) the electron donation from cytochrome b_5 to other microsomal acceptors like semidehydroascorbate[87] or other radicals or
(ii) the so-called "uncoupling" reaction which occurs when the active oxygen fails to attack the substrate and is reduced to water (Chap. 6.6).

In accordance with the latter hypothesis is the rapid reoxidation of reduced cytochrome b_5 in the presence of uncouplers like perfluoro-n-hexane[86]. Against the first hypothesis speaks that the reconstituted systems work without cytochrome b_5[88] and that the antibody against cytochrome b_5 does not inhibit the NADPH-supported monooxygenation[89]. Although a definite answer regarding the role of cytochrome b_5 does not yet seem to be at hand, its principal role as an electron donor for cytochrome P450 under special conditions is beyond doubt.

6 Mechanisms of Heme-Sulfur-Dependent Monooxygenases

The isolation of many monooxygenase systems has meanwhile greatly facilitated our understanding of their reaction mechanisms. Of major importance is the conclusion that all heme-sulfur dependent monooxygenases proceed by the same reaction cycle and that only the electron transport system may differ. Essentially five steps are involved in the reaction cycle. These will be discussed sequentially.

6.1 Substrate Binding

In almost all enzyme reactions the primary step consists in the binding of the substrate molecule to the active site of the enzyme. The more interactions of the functional groups of the substrate with the surface of the protein are involved the higher is the specificity of the complex formation and the better its affinity. This also applies for the various forms of cytochrome P450. Some monooxygenases are very specific, showing K_m-values of about 10^{-6}M and lower, and some are unspecific and have a low affinity of with K_m-values around 10^{-2}M. As a general rule most substrates are lipophilic molecules which contribute very little to the enthalpy of the binding process but lead to high entropy values which can compensate for the sometimes positive enthalpy changes. The following equilibrium and thermodynamic constants have been calculated for the two spin states (Chap. 3) of the ferric xenobiotic monooxygenase LM 2 from rabbit in the presence and absence of benzphetamine as a substrate[90] (Table 4).

In general a substrate shifts the equilibrium to the high spin form of cytochrome P450 which can be easily followed by the spectral blue shift of the Soret band from 420 to 390 nm[91]. This explains why the spectral dissociation constant, K_s, calculated for this conversion agrees with the K_m of the monooxygenase reaction[92].

Table 4. Equilibrium and thermodynamic constants of the spin equilibrium of ferric cytochrome P450 LM_2 in the absence and presence of benzphetamine (= RH)[90)]

Reaction	$K_{(20\,°C)}$	ΔH [kJ/mol]	ΔS [e. u.]	ΔG [kJ/mol]
$Fe_l^{3-} \rightleftharpoons Fe_h^{3+}$	0.08	−44.4 ± 0.8	−31 ± 1	−6.3 ± 0.8
$RHFe_l^{3+} \rightleftharpoons RHFe_h^{3+}$	0.39	−38.1 ± 0.8	−29 ± 1	−2.5 ± 0.8
$Fe_l^{3+} \underset{-RH}{\overset{+RH}{\rightleftharpoons}} RHFe_h^{3+}$	0.32	31.4 ± 2.1	41 ± 2	−18.8 ± 2.5
$Fe_h^{3+} \underset{-RH}{\overset{+RH}{\rightleftharpoons}} RHFe_l^{3+}$	0.07	26.4 ± 2.1	41 ± 2	−23.9 ± 2.5

l = Low spin.
h = High spin.

It is certain that the binding of substrates is not a largely exergonic process and therefore a thermal spin equilibrium may exist. For some cytochrome P450 species this was indeed observed[90)].

Such a change in the coordination of the metal atom is characteristic for the catalytic properties of many metalloenzymes. It requires a weak ligand binding at this coordination site which in the case of cytochrome P450 would be in accord with the proposed hydroxyl group in the sixth position of the iron[51, 59)]. The entropy effect of the substrate binding would be sufficient to shift the spin equilibrium to the five-coordinated high spin complex.

A special situation arises when an organic compound contains heteroatoms like nitrogen, sulfur, phosphines or oxygen, capable of interacting with the sixth ligand position by formation of a coordinate bond. As pointed out in chapter 3 these molecules can act as organic ligands producing low spin spectra of the types described in Table 2[93)]. The normal substrates can compete with the ligand binding by shifting the spin state back to the high spin form. This is consistent with the idea that ligands and substrates occupy the same binding site. In the case of the wellknown ligand metyrapone (2-methyl-1,2-di-3-pyridyl-1-propanone) a spin-labelled analog was used to calculate the distances between the nitroxyl radical and the ferric iron which turned out to be in agreement with the binding of one pyridine nitrogen to the iron[94)]. This also explains why these organic ligands are good inhibitors for most monooxygenases. Likewise, metyrapone is used as a diagnostic tool for adrenal functions in steroid biogenesis since it has a high affinity binding constant to the steroid-11β-monooxygenase and inhibits corticoid production even in vivo[95)].

The question arises as to the significance of substrate binding for the catalytic process. First of all, it provides a means of coupling between the active oxygen production and substrate monooxygenation, since only a close proximity of both reactants guarantees a 1 : 1 : 1 stoichiometry between reduced pyridine nucleotide, dioxygen and hydroxylated substrate. Secondly, the binding of the substrate initiates the electron flow to the heme, which is easily seen by the increased rate of pyridine nucleotide oxidation after addition of substrate to a complete monooxygenase system.

The molecular event that causes the increased electron flux to the hemoprotein is not definitely known. Most likely is a conformational change that triggers the conversion of the low spin state to the high spin ferric form. The new conformation could be a better electron acceptor and thus could speed up the reduction rate of the hemoprotein. As an alternative explanation the high-spin complex could accept the electrons more readily since its potential is about 100 mV higher than that of the low spin form which has a potential around –270 mV[42]. It is unlikely that the latter hypothesis is the only explanation, since some microsomal heme sulfur proteins are predominantly in the high spin state and are reduced more readily by substrate addition without a corresponding change in spin state (Ullrich, V., Kremers, P., unpublished). More detailed investigations on the protein-protein interaction between the reducing systems and the hemoprotein should give an answer to this problem.

6.2 Reduction of the Enzyme-Substrate Complex

It was mentioned before that the presence of substrates accelerate the reduction of cytochrome P450. The kinetics of this process can be conveniently followed under anaerobic conditions in the presence of carbon monoxide, since this ligand forms a rather stable complex which can be measured spectrophotometrically by its ab-

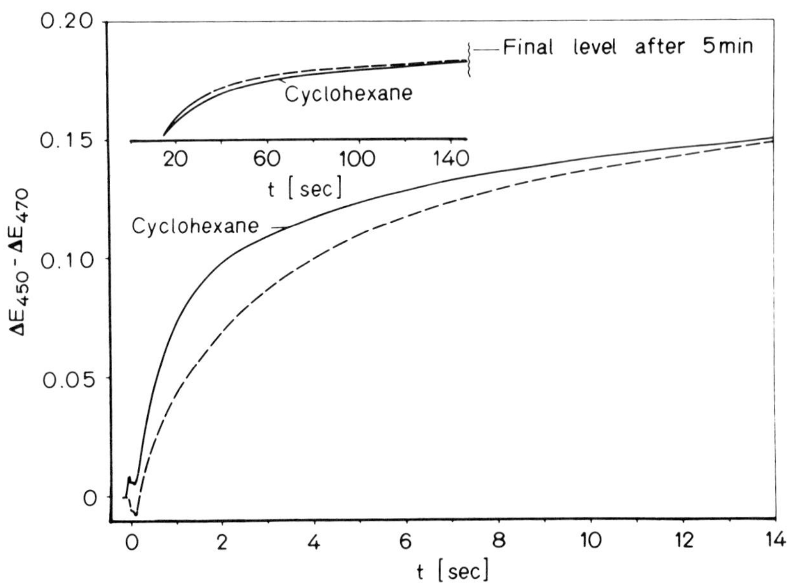

Fig. 8. Kinetics of the anaerobic reduction of microsomal cytochrome P450 by NADPH in the presence of carbon monoxide. The microsomal suspension contained 1.2 nmol cytochrome P450/ml. In both experiments 95% of the cytochrome were reduced enzymatically (dithionite = = 100%). Dashed line: no substrate added; solid line: 10^{-2} mol/l cyclohexane added. From[97]

sorption band at 450 nm. The most clear-cut results are obtained with the soluble camphor-5-exo-monooxygenase system, which shows no reduction of the hemoprotein by NADH in the absence of camphor. Liver microsomes, especially from animals pretreated with phenobarbital, show a moderate reduction rate by NADPH already without substrate, but an about 10fold enhancement is observed with substrates like cyclohexane[96, 97] (Fig. 8).

A computer analysis of the biphasic kinetics revealed a turnover number of about 66/min for the hemoprotein-cyclohexane complex, which agreed with the specific activity of cyclohexane hydroxylation based on the amount of high-spin complex formed after addition of cyclohexane[97].

This would indicate that the reduction of the enzyme-substrate complex is the rate-limiting step of the overall reaction sequence. Two other observations are in accord with this assumption. Firstly, in a reconstituted system the turnover number can be increased up to 600 when the reductase concentration has reached a saturation value (Coon, M. J., priv. commun.). Secondly, NADH can also reduce cytochrome P450 via cytochrome b_5 in liver microsomes (Chap. 5.3) and this electron transfer adds to the rate observed with NADPH which would not be the case if any other reaction except electron transfer would be rate limiting. Since NADH with some substrates often exerts a synergistic effect, it may be that in those cases the second electron transfer becomes rate limiting. This possibility will be discussed in the next chapter.

Of interest is the molecular mechanism by which the electron transfer takes place. For the bacterial camphor monooxygenase it was established by various independent methods that the putida redoxin forms a complex with the heme sulfur protein with a K_m value of about 0.5–3 μM[73]. The carboxy terminal end of putida redoxin contains a tryptophan, which seems to participate in the binding, since its cleavage by carboxypeptidase A increases the K_m for cytochrome P450 from 2–4 to 91 μM[42]. The formation of an enzyme-enzyme complex must also be postulated for the microsomal NADPH-cytochrome P450-reductase and the hemoprotein. In the membrane-bound native state of the system this interaction may occur through a lateral diffusion of the two membrane proteins[99]. The reconstituted system requires phospholipid for this interaction[78]. In accordance with this hypothesis Duppel and Ullrich[100] have reported breaks at 20 °C in the Arrhenius plots of 7-ethoxycoumarin and 4-nitroanisol 0-dealkylation, which disappeared after addition of a detergent. In contrast, Peterson et al.[101] could not find a discontinuity in the activation energy of the rapid phase of the cytochrome P450 reduction, but did find one in the slow phase. A definite answer about the organization of the monooxygenase system in the membrane seems to be difficult in view of these contradictory results, but the concept of a rigid spatial arrangement of one reductase molecule surrounded by about 8–30 cytochrome P450 molecules[102] has low probability since cytochrome b_5 also is reduced by the NADPH-cytochrome c reductase. This would indicate at least an association-dissociation equilibrium of the reductase-cytochrome P450 complexes. On the other hand after a partial denaturation of cytochrome P450-reductase[102] a fraction of the heme-sulfur proteins in microsomes is not reducible by NADPH suggesting that patches or mosaics of independent electron transport units exist in the membrane.

6.3 Formation of the Oxy-Complex

The reduced cytochrome P450 has been determined to be high spin in the case of the camphor-5-exo-monooxygenase[42]. Its coordination sphere is probably 5-coordinated, but there is no definite proof that the mercaptide ligand is still in the fifth position. Like all ferrous hemoproteins with an open 6th ligand position cytochrome P450 reacts rapidly and with a K_m value of about 10^{-6}M with dioxygen to form an oxy complex[103, 104]. Its spectral characteristics have been investigated for the camphor-binding hemoprotein, since its life-time is several minutes at room temperature. In all properties the oxy complex resembles the corresponding complexes of hemoglobin, myoglobin or tryptophan dioxygenase except for a larger temperature dependence of the quadrupole splitting in the Mößbauer spectrum of the oxy-cytochrome $P450_{CAM}$[42]. In the microsomal system the oxy form has only been observed in the difference spectrum of aerobic NADPH supplemented suspensions with benzphetamine as a substrate[105]. Its stability is probably lower and a spontaneous decomposition to O_2^- and the ferric hemoprotein occurs, since hydrogen peroxide formation by liver microsomes in the presence of NADPH and dioxygen could clearly be attributed to a cytochrome P450-dependent O_2^- formation and its subsequent disproportionation[106]. The rate of hydrogen peroxide formation is dependent on the substrate used, which can be interpreted in terms of different hemoproteins having different stabilities for their oxy complexes, or a specific influence of the substrate structure on the oxy complex or by substrate-dependent differences in the ratios of the first to the second electron transfer:

$$Fe^{3+} \xrightarrow{+e} Fe^{2+} \xrightarrow{O_2} [FeO_2] \xrightarrow{+e} \text{active oxygen complex} \tag{12}$$

It is obvious that the rates of both electron transfers determine the steady state concentration of the oxy complex. The fact that benzphetamine greatly increases the first reduction step and also shows a high steady state concentration of the oxy complex may be in accordance with the latter concept. A varying oxy complex concentration would also indicate that both electron transfer steps proceed with similar velocities but that with some substrates the first step is rate-limiting and with others the second one. This interesting hypothesis deserves further substantiation.

An interesting aspect arises when the concentration of dissolved dioxygen becomes low and finally diminishes to zero. Deprived of its natural cosubstrate dioxygen cytochrome P450 can be a strong electron donor also for other compounds and especially for reducible substrates bound closely to the iron at the active site. Nitro, nitroso, hydroxylamino, azo, N-oxide, sulfoxide or even epoxide groups can be reduced under anaerobic conditions or even in the presence of low concentrations of dioxygen if the oxidation potential of the reducible group is high enough. This also applies for polyhalogenated compounds like carbon tetrachloride which undergo subsequent one equivalent reductions yielding radicals and probably carbenes as intermediates. This mechanism is discussed again in Chaps. 6.5 and 6.6 under the aspects of complex formation and toxification reactions.

6.4 Formation of the Active Oxygen

It had early been postulated that the active oxygen complex of heme-sulfur-dependent monooxygenases is at the redox level of peroxide or the oxygen atom[48, 107]. Thus, a one-electron reduction of the oxy complex should lead to the hydroxylating species of active oxygen and this was shown experimentally in the steroid-11β-monooxygenase system[108]. Two alternative structures can be proposed for the active oxygen complex, depending whether a cleavage of the O–O bond occurs or not[109]:

$$[Fe^{III}O{-}O^{2-}]^{+} \qquad \text{or} \qquad [FeO]^{3+} \tag{13}$$

(When calculating the total charges of these complexes the two negative charges at the porphyrin ring are neglected).

Since Compound I of peroxidase has the same oxidation state and evidence for a cleavage of the O–O bond of H_2O_2 after addition to the ferric peroxidase has been presented[110], the structure of the ferryl ion, FeO, seems more likely. This has prompted studies to replace dioxygen and the two electrons by H_2O_2, but only N-demethylation reactions could be observed when liver microsomes were supplemented with H_2O_2 in the presence of substrates[111]. In constrast, organic hydroperoxides can support microsomal monooxygenations; identical or very similar substrate specificities and product pattern are observed[112]. The finding by Lichtenberger et al.[113] that iodosobenzene can act like hydroperoxides suggested that the ferric hemoprotein reacted to an intermediate $[FeO]^{3+}$ structure indicating that the active oxygen complex contains a single oxygen atom rather than a peroxo group. Ullrich et al.[59, 114] have pointed out that a mercaptide as the trans-ligand for the active oxygen could stabilize the FeO structure according to the following mesomeric forms:

$$Fe^{3+} + O \longrightarrow {-S^-}\ Fe^{V}\ O^{2-} \longleftrightarrow S^-Fe^{IV}{-}O^- \longleftrightarrow S{-}Fe^{III}=O \tag{14}$$

This may be a reason for the unique role of sulfur in the hemoprotein cytochrome P450. In order to explain the N-demethylation activity of ferric microsomal cytochrome P450 in the presence of hydrogen peroxide, Ullrich et al.[59] have formulated the following reaction:

$$S^-\ Fe^{III} + H_2O_2 \xrightarrow[-OH^-]{} S^-Fe^{V}{-}OH \longleftrightarrow S{-}Fe^{IV}{-}OH \tag{15}$$

The resulting complex would exhibit a good oxidizing but no oxygenating activity. The chloride activation of chloroperoxidase in the presence of hydrogen peroxide could also be explained by a similar mechanism[59].

$$S^-\ Fe^{IV}{-}OH + Cl^- + RH \longrightarrow S^-\ Fe^{3+} + RCl + H_2O \tag{16}$$

An experimental approach to study directly the structure of the active oxygen at cytochrome P450 has been unsuccessful so far, probably because of the short life-time of the complex. By rapid spectrophotometry Guengerich et al.[115] have detected two intermediates in the reaction of dioxygen with the reduced heme sulfur protein LM 2 from rabbit liver of which one was assigned to the oxy complex and the second to the active oxygen species. In view of the fascinating catalytic properties of this complex it would be desirable to know more about its electronic and chemical properties.

6.5 Product Formation

From the studies with ligand-type inhibitors and their competition with substrates it was concluded that the substrate binds near the heme iron, so that a close proximity to the active oxygen is guaranteed. No indication for an activation of the substrate molecule exists which would also be unlikely in view of the low free enthalpy of the binding process[90]. All products formed can be explained by an insertion of the oxygen atom into the various positions of the substrates. Whereas the specific monooxygenases form only one main product, the microsomal monooxygenases react with almost any lipophilic organic compound and lead to a variety of products. The possible pathways, which all have been verified with many compounds, are summarized in the following scheme:

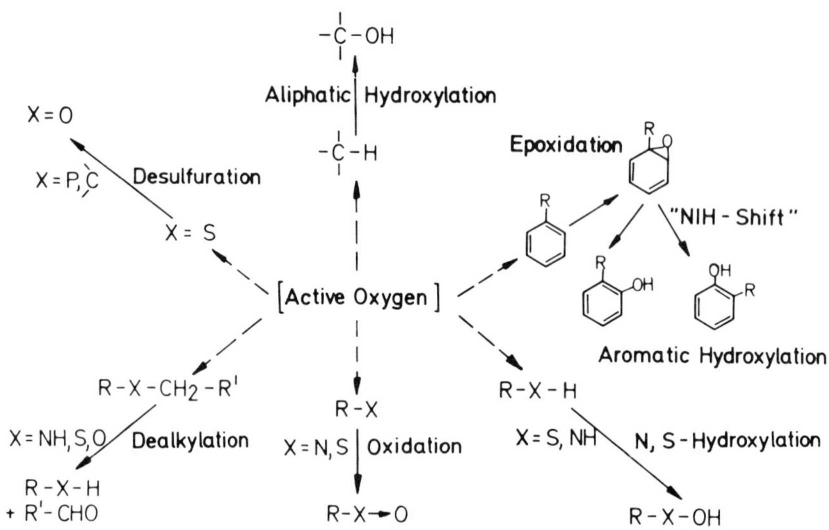

Fig. 9. Scheme of cytochrome P450-catalyzed reactions

In some cases an insertion of the oxygen atom into a substrate leads to the formation of an unstable intermediate which stabilizes by electrophilic attack at molecules of the surrounding media; e. g. semi-acetals formed by an insertion into the C–H bonds of α-carbons at heteroatoms undergo heterolytic cleavage yielding the corresponding aldehydes. If with epoxides, arene oxides or hydroxylamines the stabilization occurs by reaction with nucleophilic centers at proteins or nucleic acids, a covalent binding will result which may be the primary event leading to toxic reactions. Because of the importance of such toxifications in drug and environmental research Chap. 6.6 will be devoted to this problem.

In most cases the products resulting from the attack of the active oxygen complex are stable alcohols, phenols or epoxides. They either fulfil a certain function in cell metabolism or regulation or can be excreted directly or after conjugation with sulfate, glucuronic acid or glutathione. Each substrate has a characteristic quantitative pattern of metabolites, which may however differ with the source of microsomes. Typical examples are given in Table 5.

Characteristic for the chemical properties of the active oxygen complex are the high ratios of attack at tertiary/secondary and secondary/primary CH bonds. Aromatic monosubstituted compounds with first order substituents are hydroxylated mainly in 2- and 4-position, supporting a strongly electrophilic behavior of the active oxygen. The following experiment (Nastainczyk, Mansuy, Ullrich, unpubl.) clearly demonstrates the reactivity and unspecificity of microsomal monooxygenases: When the substrate 7-ethoxycoumarin was incubated with liver microsomes from 3-methylcholanthrene pretreated rats the reaction yields to about 95% the O-dealkylated product 7-hydroxycoumarin and only to 5% the 6-hydroxy-7-ethoxycoumarin. If now 7-methoxycoumarin was incubated under the same conditions the percent of

Table 5. Quantitative patterns of metabolites from substrates of the microsomal monooxygenase system

Substrate	Source of microsomes	Products	%	Ref.
Acetanilide	Rat, benzpyrene	4-OH-acetanilide	85	48)
	Induced	3-OH-acetanilide	3	
		2-OH-acetanilide	12	
Toluene	Rat, phenobarbital	4-OH-toluene	28	48)
	Induced	3-OH-toluene	13	
		2-OH-toluene	59	
Naphthalene	Rat, benzpyrene	1+2-Naphthol	20	48)
	Induced	1,2-Dihydrodiol	80	
n-Heptane	Rat, control	Heptane-ol-1	9.5	116)
		Heptane-ol-2	73.8	
		Heptane-ol-3	11.1	
		Heptane-ol-4	5.6	
Methylbutane	Rat, control	2-methyl-butanol-1		117)
		+3-methyl-butanol-1	6.7	
		3-methyl-butanol-1	23.8	
		2-methyl-butanol-2	69.5	
Methylcyclo-hexane (MeCH)	Rat, control	1-OH-MeCH	23.1	117)
		2-OH-MeCH	8.8	
		3-OH-MeCH	44.7	
		4-OH-MeCH	22.7	
		ω-OH-MeCH	0.7	

dealkylation and ring hydroxylation changed to 55% and 45%, respectively. Obviously the higher activation energy required for the attack of the primary CH bond in the methoxy derivative favored the ring hydroxylation. Because of the same reasons microsomal monooxygenations exhibit a large kinetic isotope effect in the order of 10–12, when two exactly equivalent CH and CD bonds are compared in one substrate molecule[118–120]. This would be in accordance with a direct cleavage of the CH bond in the transition state of the reaction. A radical abstraction would be agreement with such a large isotope effect[120]. In contrast, no primary isotope effect was observed when cyclohexane and dodeca-deutero cyclohexane were used as substrates[92]. This can easily be explained by a common rate-limiting step for both substrates consisting of the electron transfer to the hemoprotein. To a radical abstraction mechanism also points the fact that not a complete retention of the configuration at the hydroxylated carbon atom is observed[120].

The most likely sequence of events leading to product formation could thus be described as follows:

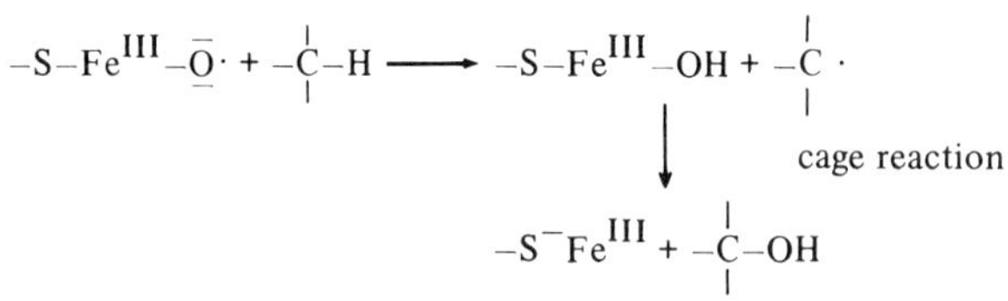

(17)

The essential step would consist in the subtraction of a hydrogen atom from the CH bond, whereas the subsequent hydroxylation is probably a cage reaction and extremely fast. It therefore may be of minor importance whether the hydroxylation proceeds by a homolytic cleavage of the Fe–OH bond or by a previous oxidation of the C-radical to a carbonium ion with simultaneous addition of the hydroxyl ion. Since the retention of the carbon atom configuration is largely retained, the lifetime of this intermediate transition state must be extremely short.

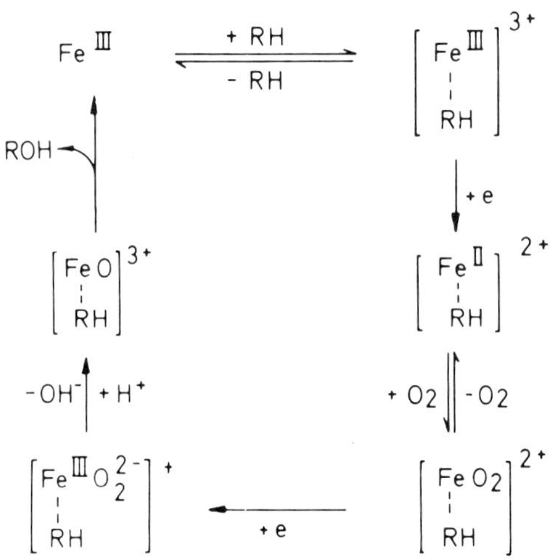

Fig. 10. Reaction scheme postulated for cytochrome P450

As a last step in the reaction sequence the product is released from the active site, which is facilitated by the increased hydrophilic properties of the oxygenated product. After the product has left the active site the hemoprotein is converted back into its low spin state and can enter a second cycle. The complete sequence of reactions can be summarized as in Fig. 10.

In some cases the product may not readily be released, but may form a ligand complex with the ferric hemoprotein or after its reduction also with the ferrous cytochrome P450. In principle this can occur with alcohols or phenols, but these heme-hydroxyl complexes are very unstable and can be easily displaced by an excess of substrate.

This is different when free amines are formed as a result of an N-demethylation reaction. Free amines bind rather strongly to the oxidized and reduced hemoprotein and thus may cause product inhibition[121]. Very stable complexes of different nature with the reduced hemoprotein have been reported after incubation of certain N-alkylated amines of the amphetamine type with liver microsomes[122]. Recent studies on the mechanism of formation of these complexes support a mechanism of N-hydroxylation leading to aliphatic nitroso compounds which form tight complexes with the ferrous heme-sulfur protein[123, 124].

Another type of complexes is formed with benzodioxoles[125]. Monooxygenation and subsequent reduction of liver microsomes anaerobically by NADPH or by sodium dithionite give rise to the following spectra[83] (Fig. 11). These complexes are very tight and resistant to exchange by other ligands. They inhibit the monooxygenation of other substrates, but this inhibition can be released by irradation with light. The photochemical action spectrum resembles the spectrum of the reduced hemoprotein-benzodioxole complex, which indicates that this complex may contain a Fe–C bond,

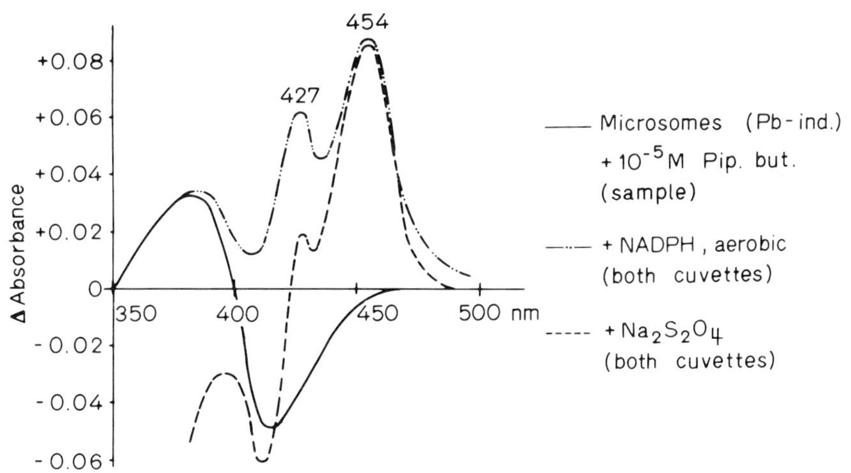

Fig. 11. Difference spectra of liver microsomes with piperonyl butoxide. From[83]

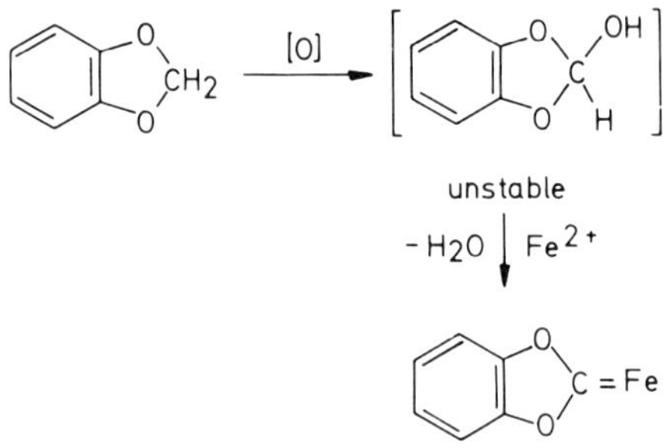

Fig. 12. Postulated mechanism of complex formation of benzodioxole with liver microsomal cytochrome P450

which is known to be sensitive to irradiation[83]. We therefore have postulated the following mechanism involving the formation of a carbene complex (Fig. 12).

Such carbene complexes may also be formed reductively by cytochrome P450 of liver microsomes from geminal polyhalogenated hydrocarbons. In the case of halothane a complex with a Soret absorption band at 470 nm is formed which was explained by a carbene formation according to the equations given in Fig. 13[126]. By an independent method using 1,1,1-trifluorodiazoethane the same difference spectrum was obtained which strongly suggests the carbene nature of the halothane ligand.

With polyhalogenated methanes similar ligand spectra under reducing conditions could be observed. As a product of this reductive pathway chloroform[127] and carbon monoxide[128] were detected. Since with $^{14}CCl_4$ the resulting CO was found to be labeled the intermediate formation of a dichlorocarbene seems likely (Fig. 14).

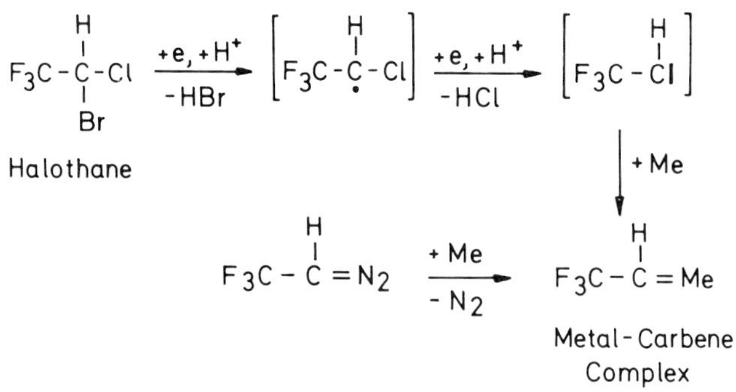

Fig. 13. Proposed mechanism of carbene complex formation with liver microsomal cytochrome P450 and halothane

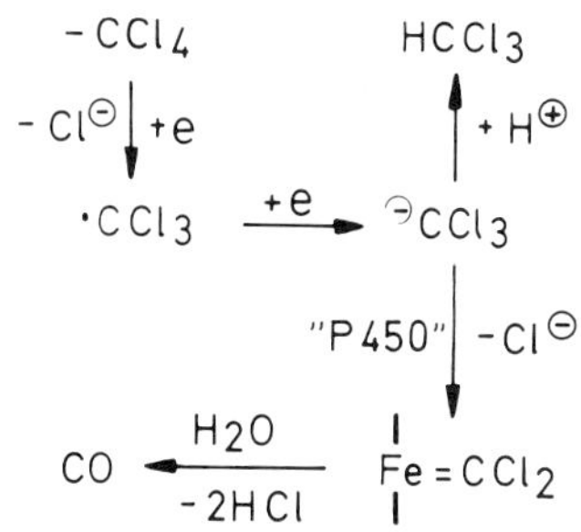

Fig. 14. Reductive dehalogenation of carbon tetrachloride by liver microsomal cytochrome P450 under carbene complex formation

The reductive dehalogenation by liver microsomal cytochrome P450 is a reaction competing rather effectively with the dioxygen reduction, especially when the oxygen pressure is low[129]. It is interesting that it also proceeds in full analogy to the dioxygen activation:

$$Fe^{2+} + O_2 \longrightarrow [FeO_2] \xrightarrow[+H^+]{+e} [Fe{=}O] + OH^- \qquad [FeO_2] \downarrow \cdot O_2^- \quad \text{(oxenoid complex)} \tag{18}$$

$$Fe^{2+} + CCl_4 \longrightarrow FeCCl_3 \xrightarrow{+e} [Fe{=}CCl_2] + Cl^- \qquad FeCCl_3 \downarrow \cdot CCl_3 \quad \text{(carbenoid complex)} \tag{19}$$

6.6 Reactive Intermediates

As a general rule the products of the microsomal monooxygenases are less toxic than the parent compound, since this system has developed as a inactivation mechanism for endogenous steroids and a detoxification mechanism for xenobiotics. Due to its unspecificity, however, with some drugs or foreign compounds the monooxygenation may proceed by formation of reactive intermediates. They show an electrophilic behavior in general. They may stabilize by further reaction with water or glutathione as outlined in the preceding chapter. If the electrophilic attack occurs, however, with sulfhydryl or amino groups of proteins or nucleic acids a covalent binding will result[130]. By the use of labeled substrates it has become clear that covalent binding of metabolites is a rather common event although it usually is a minor pathway of metabolism. Nevertheless, depending on the target molecule, very dramatic effects on the organism may be the consequence. There is ample evidence that mutagenesis and carcinogenesis are mainly caused by the activation of environmental chemicals by the microsomal monooxygenase system in liver, lung, skin or intestine[131]. Therefore, the toxification mechanisms for such chemicals have gained broadest interest in recent years.

Well established is the activation of polycyclic hydrocarbons to arene oxides[132]. 3,4-benzo(a)pyrene forms several arene oxides from which a secondary metabolite, the 9,10-dihydrodiol-7,8-epoxide has been proved highly carcinogenic[133]. It preferentially binds to deoxyguanosine and deoxyadenosine in DNA and by a series of further still unknown events leads to the formation of cancerous cells. The microsomal epoxide hydrase converts arene oxides and epoxides to inactive dihydrodiols and therefore competes with the covalent binding process[134].

A different activation mechanism is responsible for the carcinogenic properties of a series of nitrogen compounds. The primary activation step seems to be the formation of N-hydroxy derivatives in which the hydroxyl group either directly or after conjugation with sulfate or glucuronic acid serves as a leaving group[135]. N-acetylaminofluorene is a well investigated example of this group of chemicals.

An entirely different mechanism is involved in the toxicity of certain polyhalogenated compounds. As mentioned in Chaps. 6.3 and 6.5 two one electron reducing steps can lead to radicals and carbenes as reactive intermediates, which probably are responsible for lipid peroxidation and covalent binding to lipids and proteins followed by cell necrosis or even cancer[136].

Another class of potentially toxic compounds are catechols which may be formed either from phenols by secondary hydroxylation or by dehydrogenation of dihydrodiols resulting from enzymatic or spontaneous arene oxide hydration. The reactive species from catechols could be quinoid or semiquinoid oxidation products[137, 138].

Many of the toxification pathways have merely been postulated on the basis of covalently bound labeled metabolites without establishing the chemical mechanism. This seems to be an interesting challenge for chemists and toxicologists.

6.7 Uncoupling Reactions

The normal reaction cycle converts one mole of substrate under consumption of stoichiometric amounts of reduced pyridine nucleotide and dioxygen to one mole of product and water. Specific monooxygenases usually show this stoichiometry but in liver microsomes hydrogen peroxide can be formed (Chap. 6.3) which increases the relative NADPH and O_2 consumption.

In the absence of exogenous substrates liver microsomes oxidize NADPH almost exclusively to hydrogen peroxide, due to an autoxidation of the cytochrome P450 and to a minor part also the NADPH-cytochrome P450 reductase[106]. A special situation arises when a lipophilic organic compound triggers the reduction, followed by oxygenation and active oxygen complex formation, at the cytochrome but fails to incorporate the oxygen atom because of chemical or sterical reasons. Chemical reasons apply when no functional group can accept the oxygen atom as with aliphatic perfluorinated compounds[86, 139]. The C–F bond is too strong to be cleaved and no products are derived. Sterical reasons may apply if the compound binds tightly to the active site but has no hydroxylatable position reaching the catalytic site. Androstendione binding to steroid-21-monooxygenase of adrenal microsomes would be an example for this kind of "dead-end" inhibitors[140]. For bacterial flavin-dependent monooxygenases this phenomenon has first been described as "uncoupling", which later has been used also for microsomal monooxygenases[139]. Surprisingly, the flavin-dependent monooxygenases yielded hydrogen peroxide as a product of uncoupling[141], whereas with perfluoron-n-hexane as an uncoupler in the microsomal system no hydrogen peroxide was found (Kuthan, H., Ullrich, V., unpubl.), but a stoichiometry of two NADPH per one dioxygen molecule. Thus, the active oxygen complex was further reduced to water and not liberated as hydrogen peroxide, which is an indication of a different mechanism of oxygen activation in flavin-dependent monooxygenases compared with heme-sulfur monooxygenases. It is also in agreement with an $[FeO]^{3+}$ structure of the cytochrome P450-active oxygen complex, which is unlikely to attack water under formation of hydrogen peroxide. It was interesting to find that uncouplers of the microsomal system cause a more rapid co-oxidation of NADH[86]. Evidence was presented that NADH via cytochrome b_5 reduces the uncoupled active oxygen complex faster than NADPH via cytochrome P450 reductase and thus saves electrons needed for the monooxygenase reaction[86]. This phenomenon can explain, at least in part, the more than additive stimulation of the monooxygenase reactions ("synergistic effect"[84]) in the presence of both pyridine nucleotides.

It seems wise to differentiate the uncoupling reaction from the earlier described autoxidation of reduced cytochrome P450, which forms hydrogen peroxide via O_2^- radicals as the reduction product of dioxygen. Both reactions, however, can occur in vivo and are physiological side reactions due to the unspecificity of the microsomal monooxygenase system.

The origin of the side-reactions in the liver microsomal monooxygenase system is summarized in Fig. 15.

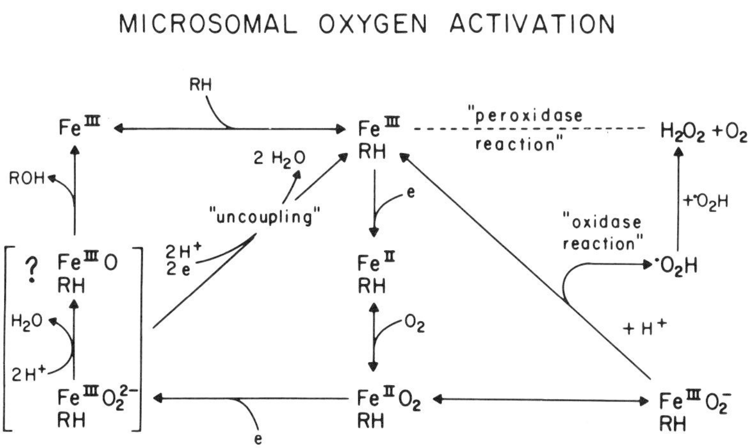

Fig. 15. Complete scheme of dioxygen activation by microsomal cytochrome P450

7 Regulation of Enzyme Activity

The description of cytochrome P450 dependent monooxygenases would be incomplete without mentioning the regulation of the enzyme activity in the cell. From the mechanism of action it is evident that the monooxygenase activity is directly proportional to the concentration of the heme-sulfur protein. In addition, however, since the electron transfer is rate limiting in all systems, the concentrations of the electron transport proteins are a second parameter. A higher activity of the iron-sulfur proteins or the flavoproteins, respectively, will increase the turnover of the hemoprotein and hence increase the monooxygenase activity.

Details of the regulatory mechanisms are not yet known, but the expression of the bacterial camphor-5-exo-monooxygenase seems to be under substrate control. A variety of closely related camphor derivatives can also act as inducers[42]. The genes for the monooxygenase are carried by plasmids and one could speculate that a depressor protein is synthetized which is inactivated in the presence of the inducer.

Much more complicated are the regulatory mechanisms involved in steroid biosynthesis. The conversion of cholesterol to glucocorticoids, mineralocorticoids, androgens or estrogens is under hormonal control and often mediated by cyclo AMP[142]. Specific receptors and feed-back control are further factors in the regulation of these important monooxygenases, without which no higher organism can survive.

Different again is the regulation of the xenobiotic monooxygenase system in higher organisms. The steady state levels of the various multiple forms of the heme-sulfur proteins seem to be controlled by the influence of the xenobiotics on the biosynthesis and degradation of the hemoproteins. When administered to the organism some compounds cause the increase of the cytochrome P450 level and to a lesser extent also of the specific content of the microsomal cytochrome P450 reductase. The increase of the cytochrome P450 does not occur uniformly for all forms but some compounds specifically induce only one or two forms. Such specific inducers are 3-methylcholanthrene or 3,4-benzo(a)pyrene[143], whereas phenobarbital[144] or polychlorinated biphenyls[145] are less specific and induce several cytochrome P450 species. The efficiency of such inducers can be demonstrated very clearly for the monooxygenase system in the intestine. The activity present in normal animals can be almost completely suppressed after feeding a diet for 3 or 4 days which was exhaustively extracted with n-hexane to remove all potentially inducing agents (Ullrich, V., unpublished). This indicates that only exogenous compounds present in the environment are inducers of this system. The liver still contains about half of its original activity after this procedure suggesting that for this organ also endogenous inducers like steroids or bile acid precursors exist. The molecular mechanism of induction by xenobiotics is also not fully understood, but the presence of cytosolic receptors for lipophilic drugs was strongly suggested[146].

8 Other Monooxygenases

To fully evaluate the role of heme-sulfur containing monooxygenases it should be briefly discussed which other prosthetic groups exist for dioxygen activation in different monooxygenases. As already mentioned a variety of bacterial monooxygenases are known which contain flavin as the oxygen reducing factor[147]. The reduced flavin probably forms a peroxide intermediate capable of hydroxylating aromatic compounds[147]. A similar role could be played by tetrahydropteridine as a cofactor in phenylalanine hydroxylation[147]. However, more recently iron was found to be necessary for the process[148], so that an activation of dioxygen by a ferrous complex is more likely and tetrahydropteridine would merely serve as an electron donor. Iron also is involved in other monooxygenases for amino acids, like 4-hydroxy-phenylpyruvate monooxygenase or prolyl-4-monooxygenase[149]. The latter enzyme uses α-ketoglutarate as an electron donor according to the equation:

$$O_2 + \text{prolyl-peptide} + \begin{array}{c} COOH \\ | \\ CH_2 \\ | \\ CH_2 \\ | \\ CO{-}COOH \end{array} \xrightarrow{Fe^{2+}} \begin{array}{c} COO \\ | \\ CH_2 \\ | \\ CH_2 \\ | \\ COOH \end{array} \begin{array}{l} + CO_2 \\ \\ + \text{4-OH-prolyl-peptide} \end{array} \qquad (20)$$

Other enzymes of the same type hydroxylate lysyl residues in a peptide chain, γ-butyrobetaine, thymidine in the 7-position and pyrimidine deoxyribonucleosides in the 2′ position[149]. Hayaishi[150] uses the term "dioxygenase" for these enzymes, although only one oxygen atom is introduced into the substrate and the other is used for the decarboxylation and appears in the succinate. Since this decarboxylation occurs with a favorable free energy it drives the reaction towards hydroxylation. These enzymes are therefore analogous to the water-forming monooxygenases and hence should be named correspondingly.

Another type of iron-containing monooxygenase was first described by Bernhardt et al.[151] and contains a two iron-two-acid-labile-sulfur cluster. It was isolated from bacteria and catalyzes the O-demethylation of 4-methoxybenzoate[151]. The corresponding electron transport chain involves NADH, a flavoprotein and a second iron-sulfur protein[152]. It seems that many more bacterial monooxygenases belong to this type rather than to the heme-sulfur-containing category.

Finally, the group of copper-containing monooxygenases should be mentioned. The most wellknown example is dopamine-β-monooxygenase. Without doubt Cu^+ is the dioxygen activating component[153], but two copper ions seem to be necessary[154]. Ascorbate serves as an electron donor. A second example of the copper-type is phenol-ortho-monooxygenase, which was one of the first monooxygenases studied[3, 155]. The peculiar property of this enzyme is that the product of the reaction, a diphenol, serves as a reductant for the cupric ions formed during the reaction. This short survey of other monooxygenases should point out that nature has developed other dioxygen activating mechanisms, although most of them involve iron and all require a two electron reduction of dioxygen.

9 Model Systems for Monooxygenases

The ease and specificity by which monooxygenases catalyze hydroxylations or epoxidations of organic compounds has fascinated chemists from the beginning. Model systems were first developed to understand the mechanism of dioxygen activation, but a direct application for organic synthesis either in a semipreparative or even in a technical scale has always been considered.

9.1 Udenfriend Systems

The first model proposed for the drug monooxygenase system in liver microsomes was described by Udenfriend et al.[156] and since then referred to as the "Udenfriend system". It consists of ferrous iron, EDTA and ascorbate which are incubated aerobically in an aqueous buffer of pH 5–8. Aromatic compounds were hydroxylated in this mixture to phenols and the substitution pattern pointed to an electrophilic mechanism. Breslow and Lukens[157] argued that the Udenfriend system was not different from the Fenton system[158] since the autoxidation of ascorbate provides hydrogen peroxide, which by the reduction with ferrous ions forms hydroxyl radicals as hydroxylating agent:

$$\text{Ascorbate} \xrightarrow{O_2} \text{Dehydroascorbate} + H_2O_2 \qquad (21)$$

$$H_2O_2 + Fe^{2+} \longrightarrow \cdot OH + Fe^{3+} + OH^- \qquad (22)$$

Later, by studying carefully the product patterns in the Udenfriend system and in a system in which hydrogen peroxide was reduced by the Fe^{2+}/EDTA complex in the presence of ascorbate, it was postulated by Staudinger and Ullrich[159, 160] that besides hydroxyl radicals a second hydroxylation mechanisms must be present, which was characterized by a random distribution of phenolic products in contrast to the electrophilic pattern observed with OH radicals. This mechanism, called the "oxenoid" mechanism, was found in all systems consisting of autoxidizing metal ions, but hydroxyl radicals were usually the predominant hydroxylating species. Only the system stannous phosphate/dioxygen showed a random pattern of products and was clearly different from that observed with OH radicals[107] (Table 6).

The stoichiometry of these hydroxylations involved one stannous complex, one dioxygen molecule and one product. By the use of the stable oxygen isotope ^{18}O the incorporation of molecular oxygen into the substrate was demonstrated. Thus, this oxenoid mechanism fulfilled exactly the stoichiometry of the monooxygenases, but did not hydroxylate in an electrophilic pattern. It was concluded that the active oxygen in this system must be a more powerful oxidant than in cytochrome P450-dependent reactions, but at the same time the existence of this mechanism demonstrated that a two-electron reduction of dioxygen by a metal complex could form a potent hydroxylating oxenoid species.

All further attempts to find a real model system with electrophilic properties of the oxenoid species proved unsuccessful. As a closest approach to the structure

Table 6. Hydroxylation pattern of organic compounds in a modified Fenton system and in the stannous phosphate-dioxygen system[107)]

Compound	Product	% of total products	
		Fe^{2+} EDTA/ascorbate/ H_2O_2/N_2	$Sn^{2+}/HPO_4^{2-}/O_2$
Acetanilide	2-OH	46,5	34
	3-OH	2,5	44
	4–OH	51,0	22
Anisol	2-OH	66	36
	3-OH	0	47
	4-OH	34	17
Toluene	2-OH	59	46
	3-OH	10	37
	4-OH	31	17
2-Methylbutane	prim. alc.	0	24
	sec. alc.	0	20
	tert. alc.	0	56

of cytochrome P450 an iron II-2-mercaptobenzoic acid complex was reacted with dioxygen in aqueous acetone[161)]. The pattern of alcoholic and phenolic products was very similar to those observed with cytochrome P450, but the migration of substituents to the adjacent ring position of the phenolic group could not be demonstrated. Recent work with five-coordinated heme-thiolate complexes also did not succeed, but this was not to be expected since an active oxygen generated in the presence of a mercaptide would immediately react and would not be available for hydroxylation reactions.

9.2 Peracids

A completely different approach started out from a suggested peroxo structure of the active oxygen and used trifluoro peracetic acid as an oxidizing agent[102, 163)]. Indeed, aliphatic and aromatic hydroxylations with strongly electrophilic product patterns were observed[164)]. A migration of substituents also occurred but the H/D kinetic isotope effect was only 2–3, compared to 10–12 in the enzymatic reaction. This was in agreement with the postulated polar reaction mechanism of the peracid[165)] in contrast to the radical hydrogen abstraction assumed for the heme-sulfur-oxenoid complex. So again, the peracids can be considered as models for an oxenoid mechanism. However, a peracid structure is not identical with the active oxygen complex, although such a hypothesis has been put forward by Hamilton[166)].

A more promising approach could be the interaction of ferric heme-sulfur complexes with oxene donors like 3-chloroperacetic acid, cumene hydroperoxide or iodosobenzene, since cytochrome P450 seems to react with these compounds by formation of the active oxygen species. Attempts in this direction have been made, but so far were unsuccessful because of the rapid oxidation of the mercaptides by the oxene donors (Ullrich, V., unpubl.).

This points to an important function of the heme and the protein in cytochrome P450 monooxygenases: the mercaptide as the fifth ligand has to be completely separated from the active oxygen, otherwise a rapid oxidation of this ligand would take place. In terms of this property of the enzyme, which will be also a critical requirement for a true model system, more sophisticated heme sulfur complexes with shielded mercaptide ligands have to be synthesized.

10 Outlook

Much progress has been made in the last ten years to elucidate the structure and mechanism of cytochrome P450-dependent monooxygenases. Much of it is indirect evidence but direct proof will require the X-ray structure of one hemoprotein, which probably will be the protein from Ps. putida. The most interesting feature still is and will be the oxygen activation process for which a very likely mechanism has been proposed, but also here a direct characterization of the active oxygen complex is necessary. Very often enzyme mechanism are merely of academic concern but in this case pharmacologists and toxicologists, as well as chemists, will profit from any progress made in this respect.

11 References

1. Warburg, O.: Schwermetalle als Wirkgruppen von Fermenten, 2nd Edit.. Freiburg: Cantor 1949
2. Wieland, H.: Über den Verlauf der Oxydationsvorgänge. Stuttgart: Enke-Verlag 1933
3. Mason, H. S., Fowlks, W., Peterson, J.: J. Amer. Chem. Soc. *77*, 2914 (1955)
4. Hayaishi, O., Katagiri, M., Rothberg, S.: J. Amer. Chem. Soc. *77*, 5450 (1955)
5. Hayaishi, O. in Hayaishi: Oxygenases. New York: Academic Press 1962
6. Mason, H. S.: Adv. Enzymol. *19*, 79 (1957)
7. Hayaishi, O. in O. Hayaishi: Molecular mechanisms of oxygen activation. New York, London: Academic Press 1974
8. Jung, C., Ristau, O.: Pharmazie *33*, 329 (1978)
9. Garfinkel, D.: Arch. Biochem. Biophys. *71*, 493 (1958)
10. Klingenberg, M., Arch. Biochem. Biophys. *75*, 376 (1958)
11. Mitoma, C., Posner, H. S. Reitz, H. C., Udenfriend, S.: Arch. Biochem. Biophys. *61*, 431 (1956)
12. Ryan, K. J., Engel, L. L.: J. biol. Chem. *225*, 103 (1957)
13. Estabrook, R. W., Cooper, D. Y., Rosenthal, O.: Biochem. Z. *338*, 741 (1963)
14. Omura, T., Sato, R.: J. biol. Chem. *239*, 2370 (1964)
15. Omura, T., Sato, R.: J. biol. Chem. *239*, 2379 (1964)
16. Omura, T., Sanders, E., Estabrook, R. W., Cooper, D. Y., Rosenthal, O.: Arch. Biochem. Biophys. *117*, 660 (1966)
17. Cooper, D. Y., Levin, S. S., Narasimhulu, S., Rosenthal, O., Estabrook, R. W.: Science *147*, 400 (1965)

18. Diehl, H., Capalna, S., Ullrich, V.: FEBS-Letters *4,* 99 (1969)
19. Ahokas, J. T., Pelkonen, O., Karki, N. T.: Biochem. Biophys. Res. Commun. *63,* 635 (1975)
20. Yawetz, A., Agosin, M., Perry, A. S.: Comp. Biochem. Physiol. *59,* 45 (1978)
21. Gallo, M., Bertrand, J. C., Azoulay, E.: FEBS-Letters *19,* 45 (1971)
22. Lebault, J. M., Lode, E. T., Coon, M. J.: Biochem. Biophys. Res. Commun. *42,* 413 (1971)
23. Murphy, P. J., West, C. A.: Arch. Biochem. Biophys. *133,* 395 (1969)
24. Hedegaard, J., Gunsalus, I. C.: J. biol. Chem. *240,* 4038 (1965)
25. Boyd, G. S.: Biochem. J., *115,* 24 P (1969)
26. Mason, H. S., North, J. C., Vanneste, M.: Fed. Proc., Fed. Amer. Soc. Exp. Biol. *24,* 1172 (1965)
27. Yu, C. A., Gunsalus, I. C., Katagiri, M., Suhara, K., Takemori, S.: J. biol. Chem. *249,* 94 (1974)
28. Berg, A., Carlstrom, K., Gustafsson, J.-A., Ingelman-Sundberg, M.: Biochem. Biophys. Res. Commun. *66,* 1414–1424 (1975)
29. J. P. Ferris, L. H., McDonals, L. H., Patrie, M. A., Martin, M. A.: Arch. Biochem. Biophys. *175,* 443–495 (1976)
30. Breskvar, K., Hudnik-Plevnik, T.: Biochem. Biophys. Res. Commun. *74,* 1192 (1977)
31. Meehan, T. D., Coscia, C. J.: Biochem. Biophys. Res. Commun. *53,* 1043 (1973)
32. Wiseman, A., (ed.): Topics in Enzyme and Fermentation, Biotechnology I, 172 (1977)
33. Schonbrod, R. D., Terriere, L. C.: Biochem. Biophys. Res. Commun. *64,* 829 (1975)
34. Bollenbacher, W. E., Smith, S. L., Wielgus, J. J., Gilbert, L. I.: Nature *268,* 660 (1977)
35. Sweat, M. L.: J. Amer. Chem. Soc. *73,* 4056 (1951)
36. Ullrich, V.: Angew. Chemie, intern. edit. *11,* 701 (1972)
37. Tomeszewski, J. E., Jerina, D. M., Daly, J. W.: Ann. Reports Medicin. Chem. *9,* 290 (1974)
38. Atsuka, Y., Okuda, K.: J. Biol. Chem. *253,* 4653 (1978)
39. Trülzsch, D., Greim, H., Czygan, P., Hutterer, F., Schaffner, F., Popper, H., Cooper, D. Y., Rosenthal, O.: Biochemistry *12,* 76 (1973)
40. Ghazarian, J. G., Hsu, P. Y., Peterson, B. L.: Arch. Biochem. Biophys. *184,* 596 (1977)
41. Orrenius, S., Ellin, A., Jakobsson, S. V., Thor, H., Cinti, D. L., Schenkman, J. B., Estabrook, R. W.: Drug Metab. Disposition, *1,* 350 (1973)
42. Gunsalus, I. C., Meeks, J. R., Lipscomb, J. D., Debrunner, P., Münck, E., in O. Hayaishi: Molecular mechanisms of oxygen activation, p. 559. New York, London: Academic Press 1974
43. Diehl, H., Schädelin, J., Ullrich, V.: Hoppe-Seyler's Z. Physiol. Chem. *351,* 1359 (1970)
44. Peterson, J. A., Griffin, B. W.: Drug Met. Disposition *1,* 14 (1973)
45. Thomas, P. E., Lu, A. Y. H., Ryan, D., West, S. B., Kawaler, J., Levin, W.: Mol. Pharmacol. *12,* 746 (1976)
46. Alvares, A. P., Schilling, G., Levin, W., Kuntzman, R.: Biochem. Biophys. Res. Commun. *29,* 521 (1967)
47. Jefcoate, C. R. E., Gaylor, F. C.: Biochemistry *8,* 3464 (1969)
48. Ullrich, V., Staudinger, Hj., in Brodie, B. B., Gillette, J. R., (eds.): Handbook of Experimental Pharmacology, Vol. XXVIII/2, p. 251. Berlin, Heidelberg, New York: Springer 1971
49. Koch, S., Tang, S. C., Holm, R. H., Fraenkel, R. B., Ibers, J. A.: J. Amer. Chem. Soc. *97,* 916 (1975)
50. Collman, J. P., Sorrell, T. N.: J. Amer. Chem. Soc. *97,* 4133 (1975)
51. Ullrich, V., Ruf, H. H., Wende, P.: Croatica Chem. Acta *49,* 213 (1977)
52. Manson, L. K., Eaton, W. A., Sligar, S. G., Gunsalus, I. C., Ganterman, M., Connell, C. R.: J. Amer. Chem. Soc. *98,* 2672 (1976)
53. Buchler, J. W., in: Porphyrins and metalloporphyrins. Smith, K. (ed.). Amsterdam: Elsevier 1975
54. Nastainczyk, W., Ruf, H. H., Ullrich, V.: Chem. Biol. Interactions *14,* 251 (1976)
55. Stern, J. O., Peisach, J.: J. biol. Chem. *249,* 7495 (1974)
56. Sharrock, M., Munck, E., Debrunner, P. G., Marshall, V., Lipscomb, J. D., Gunsalus, I. C.: Biochemistry *12,* 258 (1973)
57. Chevion, M., Peisach, J., Blumberg, W. E.: J. Biol. Chem. *252,* 3637 (1977)

58. Philson, S. B.: Ph. D. Thesis, Univers. Illinois, Urbana, Ill. USA (1977)
59. Ullrich, V., Sakurai, H., Ruf, H. H.: Acta biol. med. german., in press
60. Peterson, J. A.: Arch. Biochem. Biophys. *144,* 678 (1971)
61. Tang, S. C., Koch, S., Papaefthymiou, G. C., Foner, S., Frankel, R. B., Ibers, J. A., Holm, R. H.: J. Amer. Chem. Soc. *98,* 2414 (1976)
62. Jung, C.: Acta biol. med. german., in press
63. Murakami, K., Mason, H. S.: J. biol. Chem. *242,* 1102 (1967)
64. Hollenberg, P. F., Hager, L. P.: J. biol. Chem. *248,* 2630 (1973)
65. Chiang, R., Makinq, R., Spomer, W. E., Hager, L. P.: Biochemistry *14,* 4166 (1975)
66. Enzyme Nomenclature, p. 35. Amsterdam: Elsevier 1972
67. Coon, M. J., Ballou, D. P., Haugen, D. A., Krezoski, S. O., Nordblom, G. D., White, R. E., in: Microsomes and drug oxidations, p. 82. Ullrich, Roots, Hildebrandt, Estabrook, Conney (eds.). Oxford: Pergamon Press 1977
68. Kimura, T.: Structure and Bonding *5,* 1 (1968)
69. Tanaka, M., Honiu, M., Yasunobu, K. T., Kimura, T.: J. biol. Chem. *248,* 1141 (1973)
70. Dunham, W. R., Palmer, G., Sands, R. H., Bearden, A. J.: Biochim. Biophys. Acta *253,* 373 (1971)
71. Estabrook, R. W., Suzuki, K., Mason, J. I., Baron, J., Taylor, W. E., Simpson, E. R., Purvis, J., McCarthy, J., in: Iron-sulfur proteins I, p. 193. Lovenberg (ed.). New York, London: Academic Press 1973
72. Katakiri, M., Ganguli, B. N., Gunsalus, I. C.: J. biol. Chem. *243,* 3543 (1968)
73. Pederson, T. C., Austin, R. H., Gunsalus, I. C., in: Microsomes and drug oxidations, p. 275. Ullrich, Roots, Hildebrandt, Estabrook, Conney (eds.) Oxford: Pergamon Press 1977
74. Pederson, J. I., Oftebro, H., Vänngard, T.: Biochem. Biophys. Res. Commun. *76,* 666 (1977)
75. Pederson, J. I., Ghazarian, J. G., Orme-Johnson, N. R., DeLuca, H. F.: J. biol. Chem. *251,* 3933 (1976)
76. Williams, C. H., Kamin, H.: J. biol. Chem. *237,* 587 (1962)
77. Iyanagi, T., Mason, H. S.: Biochemistry *12,* 2297 (1973)
78. Strobel, H. W., Lu, A. Y. H., Heidema, J., Coon, M. J.: J. biol. Chem. *245,* 4851 (1970)
79. Vermilion, J. L., Coon, M. J.: J. biol. Chem. *253,* 2694 (1978)
80. Siler-Masters, B. S.: Symposium on Isolated Drug Metabolising Enzymes, Mainz 1978
81. Strittmatter, C. F., Velick, S. F.: J. biol. Chem. *228,* 785 (1957)
82. Strittmatter, C. F., Ball, E. G.: Proc. Natl. Acad. Sci. U. S. *38,* 191 (1952)
83. Ullrich, V., in: Biological reactive intermediates, p. 65. Jollow, Kocsis, Snyder, Vainio (eds.). New York, London: Plenum Press 1975
84. Cohen, B. S., Estabrook, R. W.: Arch. Biochem. Biophys. *143,* 54 (1971)
85. Hildebrandt, A. G., Estabrook, R. W.: Arch. Biochem. Biophys. *143,* 66 (1971)
86. Staudt, H., Lichtenberger, F., Ullrich, V.: Eur. J. Biochem. *46,* 99 (1974)
87. Kersten, H., Kersten, W., Staudinger, Hj.: Biochim. Biophys. Acta *27,* 598 (1958)
88. Lu, A. Y. H., Junk, K. W., Coon, M. J.: J. biol. Chem. *244,* 3714 (1969)
89. Sasame, H. A., Mitchell, J. R., Thorgeirsson, S., Gillette, J. R.: Drug Metabolism Disposition *1,* 150 (1973)
90. Rein, H., Ristau, O.: Pharmazie *33,* 325 (1978)
91. Remmer, H., Schenkman, J., Estabrook, R. W., Sesame, H., Gillette, J. R., Narasimhulu, S., Cooper, D. Y., Rostenthal, O.: Mol. Pharmacol. *2,* 187 (1966)
92. Ullrich, V.: Hoppe-Seyler's Z. Physiol. Chem. *350,* 357 (1969)
93. Ullrich, V., Schnabel, K. H.: Drug Metabolism Disposition *1,* 176 (1973)
94. Griffin, B. W., Smith, S. M., Peterson, J. A.: Arch. Biochem. Biophys. *160,* 323 (1974)
95. Liddle, G. W., Island, D., Lance, E. M., Harris, A. P.: J. Clin. Endocrinol. Metab. *18,* 906 (1958)
96. Gigon, P. L., Gram, T. E., Gillette, J. R.: Mol. Pharmacol. *5,* 109 (1969)
97. Diehl, H., Schädelin, J., Ullrich, V.: Hoppe-Seyler's Z. Physiol. Chem. *351,* 1359 (1970)
98. Sligar, S., Gunsalus, I. C.: Proc. Nat. Acad. Sci. *73,* 1078 (1976)
99. Yang, C. S., in: Microsomes and drug oxidations, p. 9. Ullrich, Roots, Hildebrandt, Estabrook, Conney (eds.). Oxford: Pergamon Press 1977

100. Duppel, W., Ullrich, V.: Biochim. Biophys. Acta *426*, 399 (1976)
101. Peterson, J. A., Ebel, R. E., O'Keeffe, D. H., Matsubara, T., Estabrook, R. W.: J. biol. Chem. *251*, 4010 (1976)
102. Franklin, M. R., Estabrook, R. W.: Arch. Biochem. Biophys. *143*, 318 (1971)
103. I. C. Gunsalus, Conf. "P450 Structure Function", Stockholm 1970
104. Ishimura, Y., Ullrich, V., Peterson, J. A.: Biochem. Biophys. Res. Commun. *42*, 140 (1971)
105. Estabrook, R. W., Hildebrandt, A. G., Baron, J., Netter, K., Leibman, K.: Biochem. Biophys. Res. Commun. *42*, 132 (1971)
106. Kuthan, H., Tsuji, H., Graf, H., Ullrich, V., Werringloer, J., Estabrook, R. W.: FEBS Letters *91*, 343 (1978)
107. Ullrich, V., Staudinger, Hj.: Z. Naturforschg. *24b*, 583 (1969)
108. Schleyer, H., Cooper, D. Y., Levin, S. S., Rosenthal, O., in: Biological hydroxylation mechanisms, p. 187. Boyd, G. S., Smellie (eds.). New York: Academic Press 1972
109. Ullrich, V., Staudinger, Hj., in: Biochemie des Sauerstoffs, p. 229. Hess, B., Staudinger, Hj. (eds.). Berlin, Heidelberg, New York: Springer 1968
110. Schonbaum, G. R., Lo, S.: J. biol. Chem. *247*, 3353 (1972)
111. Estabrook, R. W., Werringloer, J., in: Microsomes and drug oxidaions, p. 748. Ullrich, Roots, Hildebrandt, Estabrook, Conney (eds.). Oxford: Pergamon Press 1977
112. Hrycay, E. G., O'Brian, P. J.: Arch. Biochem. Biophys. *153*, 480 (1972)
113. Lichtenberger, F., Nastainczyk, W., Ullrich, V.: Biochem. Biophys. Res. Commun. *70*, 939 (1976)
114. Ullrich, V., in: Microsomes and drug oxidations, p. 192. Ullrich, Roots, Hildebrandt, Estabrook, Conney (eds.). Oxford: Pergamon Press 1977
115. Guengerich, F. P., Ballou, D. P., Coon, M. J.: Biochem. Biophys. Res. Commun. *70*, 951 (1976)
116. Frommer, U., Ullrich, V., Staudinger, Hj., Orrenius, S.: Biochim. Biophys. Acta *280*, 487 (1972)
117. Frommer, U., Ullrich, V., Staudinger, Hj.: Hoppe-Seyler's Z. Physiol. Chem. *351*, 903 (1970)
118. Foster, A. B., Jarman, M., Stevens, J. D., Thomas, P., Westwood, J. H.: Chem. Biol. Interact. *9*, 327 (1974)
119. Hjemeland, L. M., Aronow, L., Trudell, J. R.: Biochem. Biophys. Res. Commun. *76*, 541 (1977)
120. Groves, J. T., McClusky, G. A., White, R. E., Coon, M. J.: Biochem. Biophys. Res. Commun. *81*, 154 (1978)
121. Jeffcoate, C. R. E., Gaylor, J. L., Calabrese, R. L.: Biochemistry *8*, 3455 (1969)
122. Franklin, M. R.: Xenobiotica *4*, 133 (1974)
123. Mansuy, D., Gaus, P., Chottard, J. C., Bartoli, J. F.: Eur. J. Biochem. *76*, 607 (1977)
124. Mansuy, D., Chottard, J. C., Chottard, G.: Eur. J. Biochem. *76*, 617 (1977)
125. Hodgson, E., Philpot, R. M.: Drug Met. Rev. *3*, 231 (1974)
126. Mansuy, D., Nastainczyk, W., Ullrich, V.: Naunyn-Schmiedeberg's Arch. Pharmacol. *285*, 315 (1974)
127. Uehleke, H., Hellmer, K. H., Tabarelli, S.: Xenobiotica *3*, 1 (1973)
128. Wolf, R. C., Mansuy, D., Nastainczyk, W., Deutschmann, G., Ullrich, V.: Mol. Pharmacol. *13*, 698 (1977)
129. Nastainczyk, W., Ullrich, V., Sies, H.: Biochem. Pharmacol. *27*, 387 (1978)
130. Miller, E. C., Miller, J. A.: Pharmacol. Rev. *18*, 805 (1966)
131. Miller, J. A.: Cancer Res. *30*, 559 (1970)
132. Sims, P., Groves, P. L.: Adv. Cancer Res. *20*, 165 (1974)
133. Sims, P., Grover, P. L., Swaisland, A., Pal, K., Hewer, A.: Nature (London) *252*, 326 (1974)
134. Oesch, F.: Xenobiotica *3*, 305 (1973)
135. Weisburger, J. H., Yamamoto, R. S., Williams, G. H., Grantham, P. H., Matsushima, T., Weisburger, E. K.: Cancer Res. *32*, 491 (1972)
136. Recknagel, R. O.: Pharmacol. Rev. *19*, 145 (1967)
137. Marks, F., Hecker, E.: Biochim. Biophys. Acta *187*, 250 (1969)
138. Bolt, H., Remmer, H.: Xenobiotica *2*, 489 (1972)

139. Ullrich, V., Diehl, H.: Eur. J. Biochem. *20*, 509 (1971)
140. Narasimhulu, S.: Arch. Biochem. Biophys. *147*, 384 (1971)
141. White, Stevens, R. H., Kamin, H.: Biochem. Biophys. Res. Commun. *38*, 882 (1970)
142. Koritz, S. B., Kumar, A. M.: J. biol. Chem. *245*, 152 (1970)
143. Conney, A. H., Lu, A. Y. H., Levin, W., Somogy, A., West, S., Jacobson, M., Ryan, D., Kuntzman, R.: Drug Metabolism Disposition *1*, 199 (1973)
144. Welton, A. F., O'Neal, F. O., Choney, L. C., Aust, S. D.: J. biol. Chem. *250*, 563 (1975)
145. Alvares, A. P., in: Microsomes and drug oxidations, p. 476. Ullrich, Roots, Hildebrandt, Estabrook, Conney (eds.). Oxford: Pergamon Press 1977
146. Poland, A., Grover, E., Kende, A. S.: J. biol. Chem. *251*, 4936 (1976)
147. Massey, V., Hemmerich, P.: The Enzymes 3rd edit., Vol. XII, B, p. 191. Boyer, P. (ed.). New York, San Francisco, London: Academic Press 1975
148. Petrack, B., Sheppy, F., Fetzer, V.: J. biol. Chem. *243*, 743 (1968)
149. Abbott, H. T., Udenfriend, S., in: Molecular mechanisms of oxygen activation, p. 167. Hayaishi, O. (ed.). New York, London: Academic Press 1974
150. Hayaishi, O., in: Molecular mechanisms of oxygen activation, p. 1. New York, London: Academic Press 1974
151. Bernhardt, F. H., Staudinger, Hj., Ullrich, V.: Hoppe-Seyler's Z. Physiol. Chem. *351*, 467 (1970)
152. Bernhardt, F. H., Staudinger, Hj.: Hoppe-Seyler's Z. Physiol. Chem. *354*, 217 (1973)
153. Friedman, S., Kaufman, S.: J. biol. Chem. *240*, PC 552 (1965)
154. Friedman, S., Kaufman, S.: J. biol. Chem. *241*, 2256 (1966)
155. Vanneste, W. H., Zuberbühler, A., in: Molecular mechanisms of oxygen activation, p. 317. Hayaishi, O. (ed.). New York, London: Academic Press 1974
156. Udenfriend, S., Clark, C. T., Axelrod, J., Brodie, B. B.: J. biol. Chem. *208*, 731 (1954)
157. Breslow, R., Lukens, L. N.: J. biol. Chem. *235*, 292 (1960)
158. Loebl, H., Stein, G., Weiss, J.: Chem. Soc. (London) *1950*, 2704
159. Staudinger, Hj., Ullrich, V.: Z. Naturforschg. *19b*, 409 (1964)
160. Staudinger, Hj., Ullrich, V.: Z. Naturforschg. *19b*, 877 (1964)
161. Ullrich, V.: Z. Naturforschg. *24b*, 699 (1969)
162. Davidson, A. J., Norman, R. O. C.: J. Chem. Soc. (London) *1964*, 5404
163. Ullrich, V., Wolf, J., Amadori, E., Staudinger, Hj.: Hoppe-Seyler's Z. Physiol. Chem. *349*, 85 (1968)
164. Frommer, U., Ullrich, V.: Z. Naturforschg. *26b*, 322 (1971)
165. Ibne-Rasa, K. M., Edwards, J. O.: J. Amer. Chem. Soc. *84*, 763 (1971)
166. Hamilton, G. A., in: Molecular mechanisms of oxygen activation, p. 405. Hayaishi, O. (ed.). New York, London: Academic Press 1974

Received November 27, 1978

Aminoglycoside Antibiotics

Chemistry, Biochemistry, Structure-Activity Relationships

Jürgen Reden and Walter Dürckheimer

Hoechst Aktiengesellschaft, Postfach 80 03 20, 6230 Frankfurt (M) 80, Germany

Table of Contents

1 Introduction

Aminoglycoside antibiotics play a central role in the therapy of infectious diseases. Streptomycin, the first member of this class, was isolated from the culture filtrate of *Streptomyces griseus* by S. A. Waksman[1)] in 1944. Since then a great number of new aminoglycosides have been discovered and their structures elucidated. For many years they were considered unfavorable for therapeutic use as antibacterials, because of their oto- and nephrotoxicity and their lack of enteral absorption.

The discovery of Gentamycin[2)] and its successful utilisation for the treatment of serious gram-negative infections has contributed in the last years to increasing clinical use and to a rapid development of the chemistry and biochemistry of aminoglycoside antibiotics. Also of importance in the further exploration of this class of compounds are the new chemical and physico-chemical methods, e.g. H- and ^{13}C-NMR-spectroscopy, mass spectroscopy, X-ray analysis and the different chromatographic methods. Synthetic efforts and studies of the inactivation mechanism have provided a preliminary insight into the relationship between structure and activity.

There are already several comprehensive reviews of the chemical, biochemical and medicinal aspects of aminoglycosides[3–10a)]. This report mainly describes the progress made in chemistry and medicinal chemistry since 1970. Previous work will be discussed only when it appears necessary for better understanding.

Table 1. Therapeutically important aminoglycoside antibiotics

Chemical name	Structural type	Year of discovery	Trade name
Streptomycin	1	1944	Solvo-Strept "S"[a], Streptothenat[b]
Dihydro-streptomycin	1		Solvo-strept[a]
Neomycin B	2b	1949	Bycomycin[a], Framycetin[a], Myacyne[a]
Paromomycin	2b	1959	Gabbromycin[a], Humatin[a]
Kanamycin A	2c	1957	Resistomycin[a], Kanabristol[a], Kanamytrex
Kanamycin B	2c		Kanendomycin
Gentamycin	2c	1963	Refobacin[a], Sulmycin[a], Garamycin
Tobramycin	2c	1967	Gernebcin[a]
Sisomicin	2c	1970	Extramycin[b], Pathomycin[b]
Amikacin	2c	1970	Biklin[b], Amikin
Spectinomycin	3	1960	Stanilo[a]

[a] Rote Liste, Editio Cantor, Aulendorf/Württemberg (1975).
[b] Die Liste Pharmaindex III/76, I.M.P. Verlagsgesellschaft mbH, Neu-Isenburg.

II Basic Structure, Structure-Elucidation and Physico-Chemical Properties

Aminoglycoside antibiotics contain as a fundamental structural feature an aminocyclitol moiety, which has a glycosidic link with one or more sugars or amino-sugars. Generally the glycosidic linkages are of the α-configuration. Depending on the aminocyclitol present and the site of the linkages, the aminoglycosides can be divided into the following types:

Type 1: Streptidine-containing antibiotics

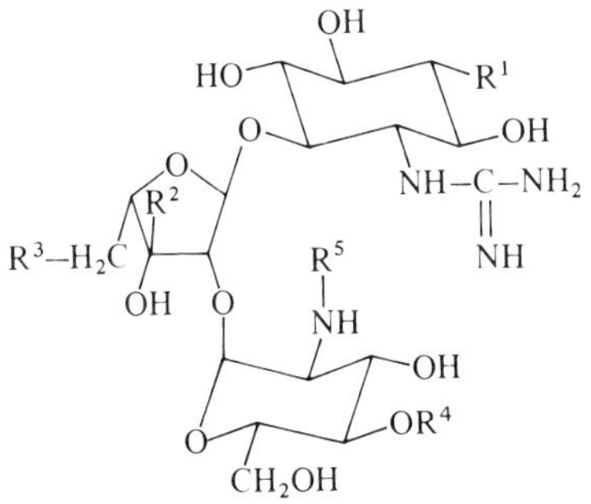

Fig. 1. Structural type 1

	R^1	R^2	R^3	R^4	R^5	Ref.
1 Streptomycin	$-NHC(=NH)NH_2$	CHO	H	H	CH_3	1)
2 N-Demethylstreptomycin	$-NHC(=NH)NH_2$	CHO	H	H	H	11)
3 Hydroxystreptomycin	$-NHC(=NH)NH_2$	CHO	OH	H	CH_3	12)
4 Mannosidostreptomycin	$-NHC(=NH)NH_2$	CHO	H	a	CH_3	13)
5 Mannosidohydroxystreptomycin	$-NHC(=NH)NH_2$	CHO	OH	a	CH_3	14)
6 Dihydrostreptomycin	$-NHC(=NH)NH_2$	CH_2OH	H	H	CH_3	15)
7 Bluensomycin	$-OCONH_2$	CH_2OH	H	H	CH_3	16)

a

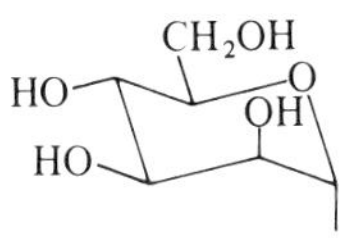

Type 2: Deoxystreptamine-containing antibiotics

Type 2a: 4-Substituted deoxystreptamine

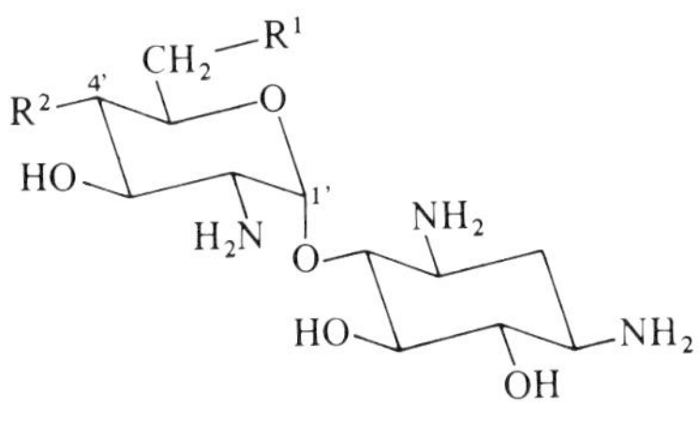

Fig. 2. Structural type 2a

	R^1	R^2	Ref.
8 Neamine	NH_2	OH	17)
9 Paromamine	OH	OH	18)
34 Seldomycin factor 2 (XK-88-2) = 4-Deoxyneamine	NH_2	H	34)

Type 2b: 4,5-Disubstituted deoxystreptamine

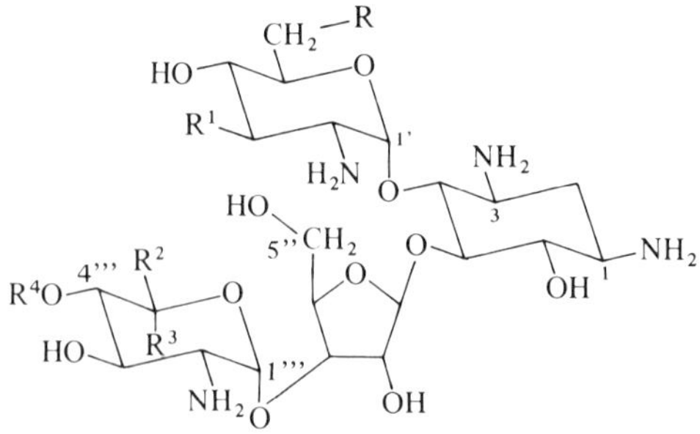

Fig. 3. Structural type 2b

	R	R^1	R^2	R^3	R^4	Ref.
10 Neomycin B	NH_2	OH	H	CH_2NH_2	H	19)
11 Neomycin C	NH_2	OH	CH_2NH_2	H	H	19)
12 Paromomycin I	OH	OH	H	CH_2NH_2	H	20, 19b)
13 Paromomycin II	OH	OH	CH_2NH_2	H	H	21, 19b)
14 Lividomycin A (Quintomycin B)	OH	H	H	CH_2NH_2	a	22)
15 Lividomycin B (Quintomycin D)	OH	H	H	CH_2NH_2	H	22)
16 Quintomycin A	OH	OH	H	CH_2NH_2	a	22)

a

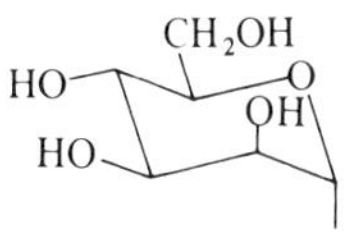

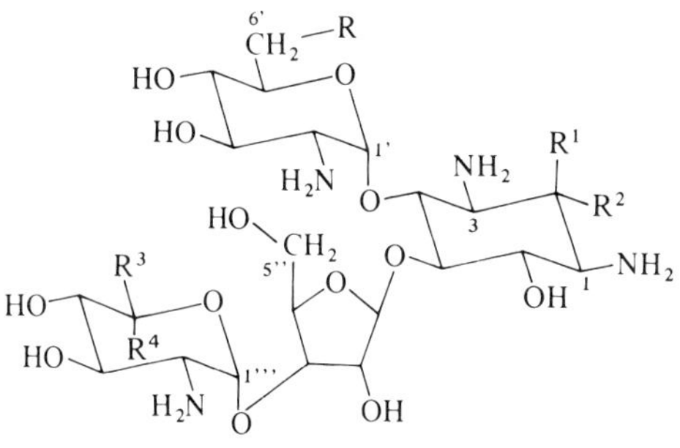

Fig. 4. Structural type 2b

	R	R^1	R^2	R^3	R^4	Ref.
17a Hybrimycin A_1	NH_2	H	OH	H	CH_2NH_2	23)
b Hybrimycin A_2	NH_2	H	OH	CH_2NH_2	H	23)
c Hybrimycin B_1	NH_2	OH	H	H	CH_2NH_2	23)
d Hybrimycin B_2	NH_2	OH	H	CH_2NH_2	H	23)
e Hybrimycin C_1	OH	H	OH	H	CH_2NH_2	23)
f Hybrimycin C_2	OH	H	OH	CH_2NH_2	H	23)

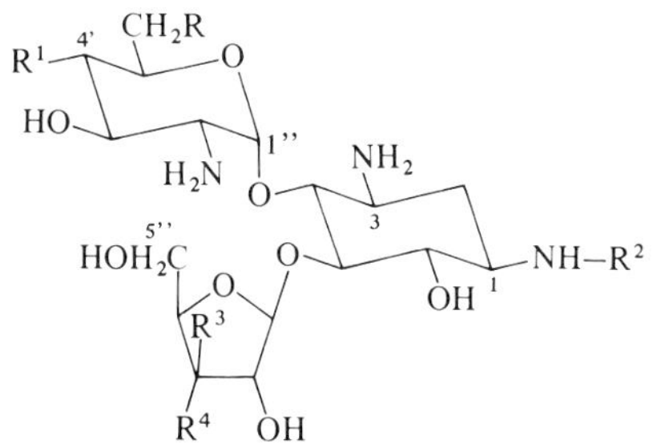

Fig. 5. Structural type 2b

	R	R^1	R^2	R^3	R^4	Ref.
18 Ribostamycin	NH_2	OH	H	H	OH	24)
19 Xylostasin	NH_2	OH	H	OH	H	25)
20 Butirosin A	NH_2	OH	a	OH	H	26)
21 Butirosin B	NH_2	OH	a	H	OH	26)
22 4'-Deoxybutirosin A	NH_2	H	a	OH	H	27)
23 4'-Deoxybutirosin B	NH_2	H	a	H	OH	27)
24 LL BM408α	OH	OH	H	H	OH	28)
25 BU-1709E_1	OH	OH	a	OH	H	29)
26 BU-1709E_2	OH	OH	a	H	OH	29)

[a] = CO−CHOH−CH_2−CH_2NH_2.

Type 2c: 4,6-Disubstituted deoxystreptamine

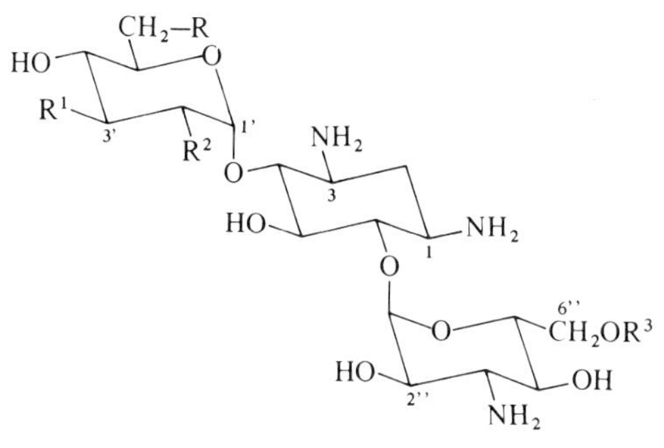

Fig. 6. Structural type 2c

	R	R^1	R^2	R^3	Ref.
27 Kanamycin A	NH_2	OH	OH	H	30)
28 Kanamycin B	NH_2	OH	NH_2	H	31)
29 Kanamycin C	OH	OH	NH_2	H	32)
30 Tobramycin (Nebramycin factor 6)	NH_2	H	NH_2	H	33)
31 Nebramycin factor 4	NH_2	OH	NH_2	$CONH_2$	33)
32 Nebramycin factor 5'	NH_2	H	NH_2	$CONH_2$	33)
59 Nebramycin factor 2	Apramycin, see Fig. 10				

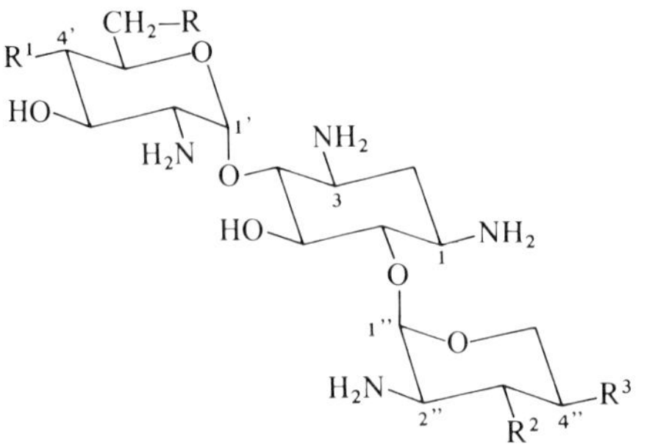

Fig. 7. Structural type 2c

	R	R^1	R^2	R^3	Ref.
33 Seldomycin factor 1 (XK-88-1)	OH	OH	OH	OH	34)
34 Seldomycin factor 2 (XK-88-2)	See structural type 2a				34)
35 Seldomycin factor 3 (XK-83-3)	NH_2	OH	OH	OH	34)
36 Seldomycin factor 5 (XK-88-5)	NH_2	H	NH_2	OCH_3	34)

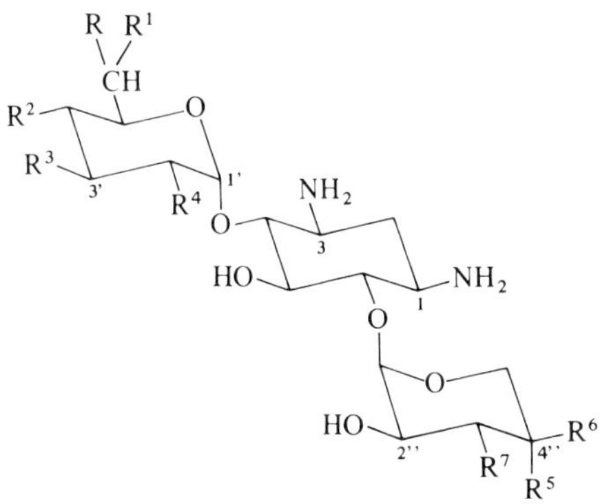

Fig. 8. Structural type 2c

	R	R^1	R^2	R^3	R^4	R^5	R^6	R^7	Ref.
37 Gentamicin C_1	CH_3	$NHCH_3$	H	H	NH_2	OH	CH_3	$NHCH_3$	35, 36)
38 Gentamicin C_2	CH_3	NH_2	H	H	NH_2	OH	CH_3	$NHCH_3$	35, 36)
39 Gentamicin C_{1a}	H	NH_2	H	H	NH_2	OH	CH_3	$NHCH_3$	35, 36)
40 Gentamicin C_{2a}[a]	CH_3	NH_2	H	H	NH_2	OH	CH_3	$NHCH_3$	36, 37b)
41 Gentamicin $C_{2\text{-}III}$[b]	H	NH_2	H	H	NH_2	OH	CH_3	$NHCH_3$	36)
42 Sagamicin (XK-62-2) Gentamicin C_{2b}	H	$NHCH_3$	H	H	NH_2	OH	CH_3	$NHCH_3$	37)

	R	R^1	R^2	R^3	R^4	R^5	R^6	R^7	Ref.
43 Gentamicin A	H	OH	OH	OH	NH_2	H	OH	$NHCH_3$	38)
44 Gentamicin A_1	H	OH	OH	OH	NH_2	OH	H	$NHCH_3$	39)
45 Gentamicin A_2	H	OH	OH	OH	NH_2	H	OH	OH	40)
46 Gentamicin A_3	H	NH_2	OH	OH	OH	OH	H	$NHCH_3$	39)
47 Gentamicin A_4	H	OH	OH	OH	NH_2	H	OH	$N(CHO)CH_3$	39)
48 Gentamicin X_2	H	OH	OH	OH	NH_2	OH	CH_3	$NHCH_3$	41)
49 Gentamicin B	H	NH_2	OH	OH	OH	OH	CH_3	$NHCH_3$	41)
50 Gentamicin B_1	CH_3	NH_2	OH	OH	NH_2	OH	CH_3	$NHCH_3$	41)
51 G 418	CH_3	OH	OH	OH	NH_2	OH	CH_3	$NHCH_3$	42)
52 JI-20A	H	NH_2	OH	OH	NH_2	OH	CH_3	$NHCH_3$	43)
53 JI-20B	CH_3	NH_2	OH	OH	NH_2	OH	CH_3	$NHCH_3$	43)

[a] Stereoisomer of C_2 at C-6′
[b] Stereoisomer of C_{1a} at C-5′
For some other gentamicin related antibiotics, produced by gentamicin C producing *Micromonospora* species in a minor amount see Ref.[44].

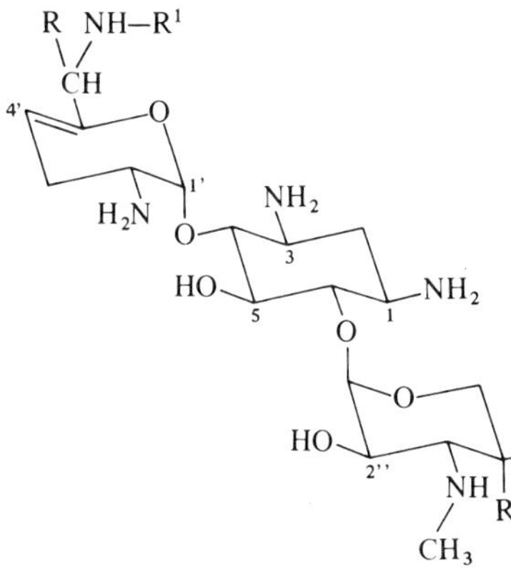

	R	R^1	R^2	R^3	Ref.
54 Sisomicin	H	H	OH	CH_3	45)
55 Verdamicin	CH_3	H	OH	CH_3	42, 46)
56 G-52(Sch17726)	H	CH_3	OH	CH_3	47)
57 66-40B	H	H	H	OH	48)
58 66-40D	H	H	OH	H	48)

Fig. 9. Structural type 2c

Type 3: Different structures

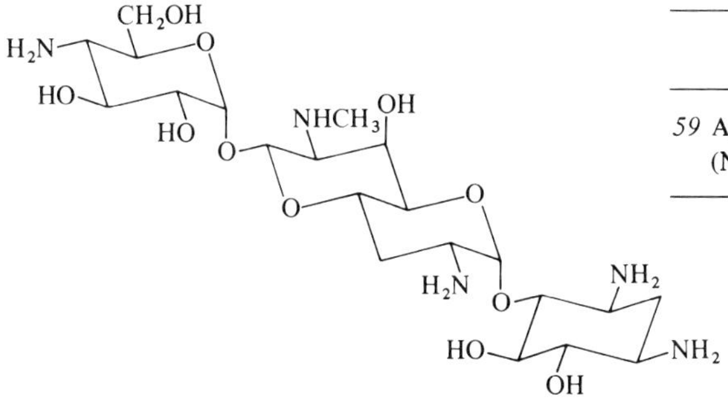

	Ref.
59 Apramycin (Nebramycin factor 2)	33, 49)

Fig. 10.

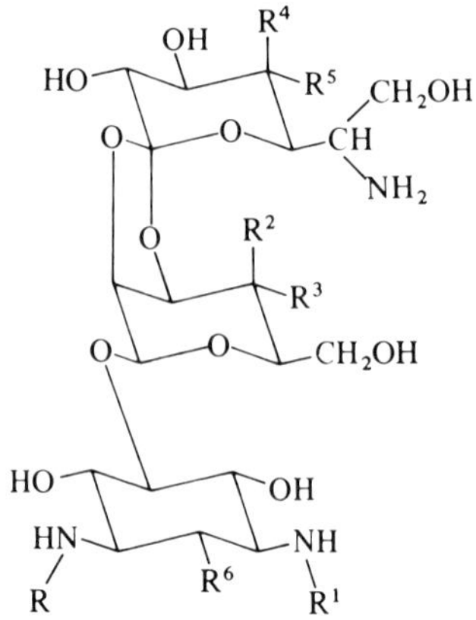

Fig. 11.

		R	R^1	R^2	R^3	R^4	R^5	R^6	Ref.
60a	Destomycin A	CH_3	H	OH	H	OH	H	H	50, 50a)
b	Destomycin B	CH_3	CH_3	H	OH	H	OH	H	50)
c	Destomycin C	CH_3	CH_3	OH	H	OH	H	H	50b)
d	Hygromycin B	H	CH_3	OH	H	OH	H	H	51)
e	SS-56 C	H	H	OH	H	OH	H	OH	52)
f	SS-56 D (A-396-I)	H	H	OH	II	OH	H	H	52)

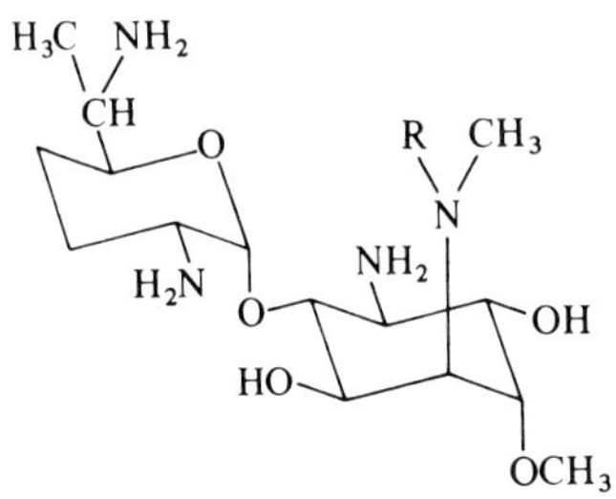

61a, c

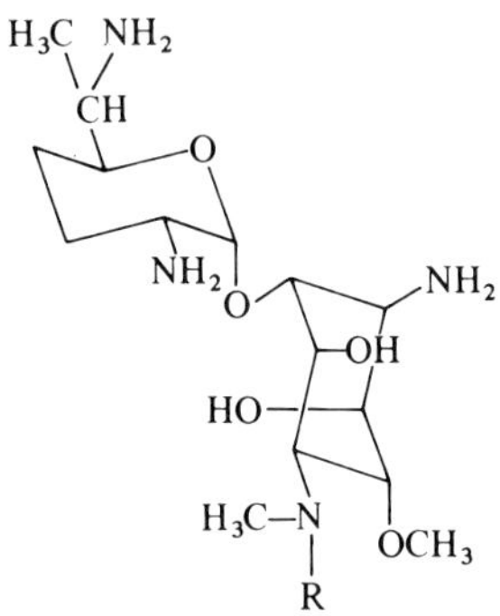

61b

Fig. 12.

		R	Ref.
61a	Fortimicin A (XK-70-1)	$-CO-CH_2NH_2$	53, 53a)
b	Fortimicin B (XK-70-2)	H	53, 53b)
c	Fortimicin C	$-CO-CH_2NH-CONH_2$	53e)

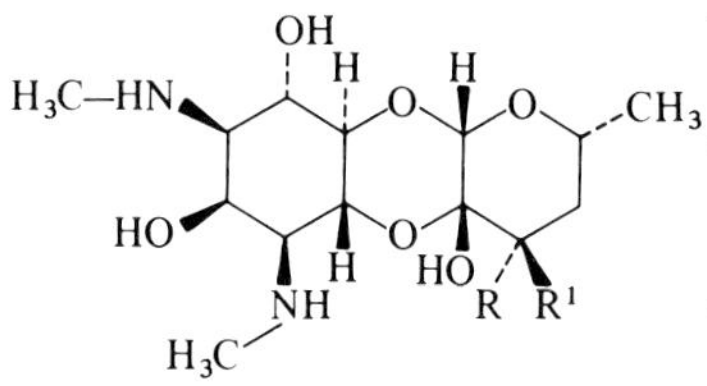

		R	R¹	Ref.
62	Spectinomycin (hydrate form)	OH	OH	[54]
62a	Dihydrospectinomycin	H	OH	[54b]

Fig. 13.

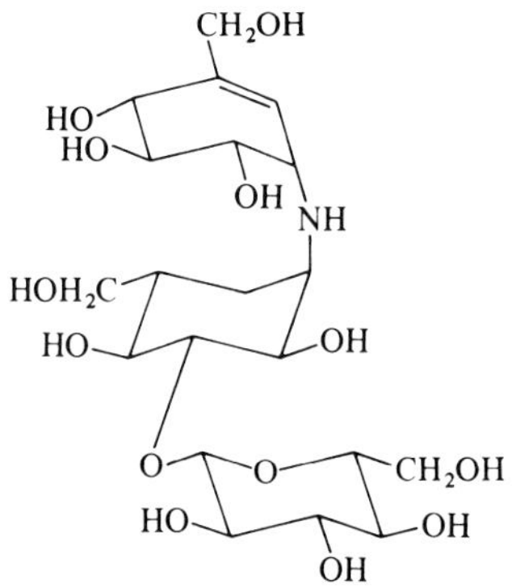

		R	Ref.
63a	Validamycin A	H	[55]
b	Validamycin E	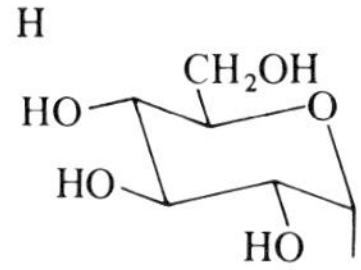	[55]

Fig. 14.

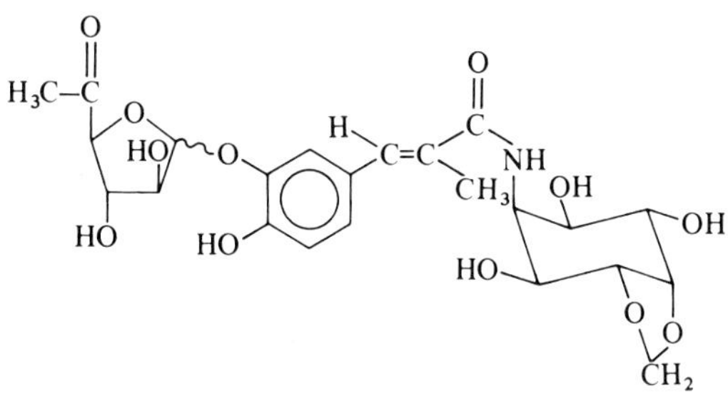

64 Hygromycin A. Ref.[56]

Fig. 15.

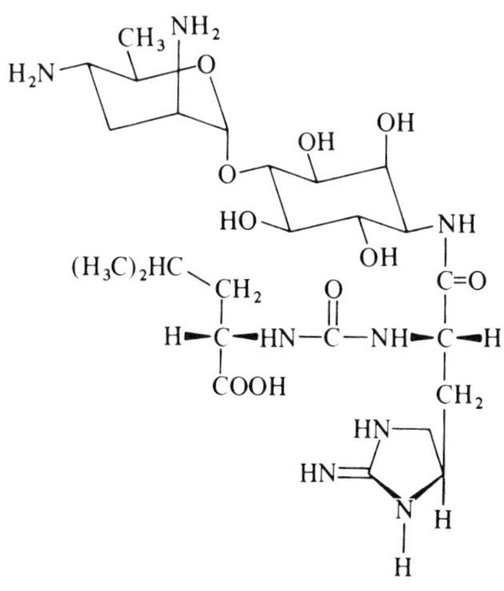

65 Minosaminomycin. Ref.[57]

Fig. 16.

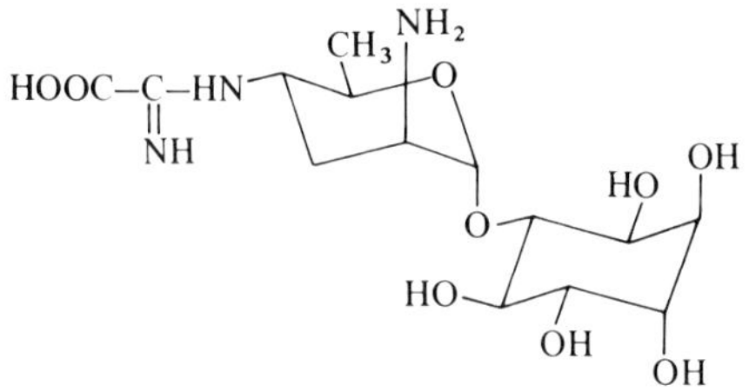

66 Kasugamycin. Ref.[58]

Fig. 17.

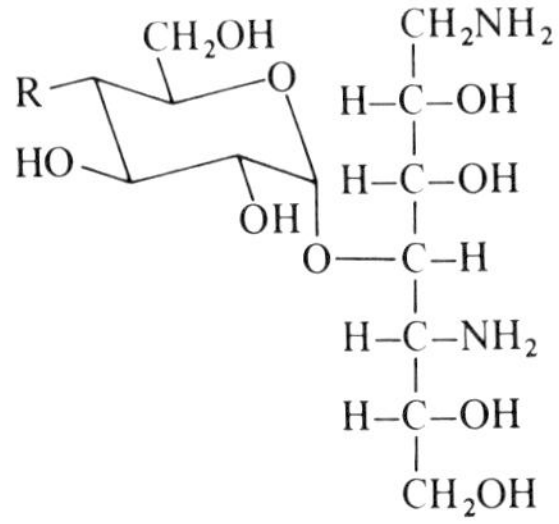

Fig. 18.

		R	Ref.
67a	Sorbistin A_1	$-NH-COCH_2CH_3$	59)
b	Sorbistin A_2	$-NH-COCH_2CH_2CH_3$	59)
c	Sorbistin B	$-NH-COCH_3$	59)
d	Sorbistin D	$-NH_2$	59)
e	Sorbistin C	$-OH$	59)
68a	AM 31α	$-NH_2$	60)
b	AM 31β	$-NH-COCH_2CH_3$	60)
c	AM 31γ	$-NH-COCH_3$	60)

Many physico-chemical properties of the aminoglycosides are similar because of close relations in the structural units of the compounds. The aminoglycosides are stable, colourless, amorphous basic compounds, readily soluble in water and slightly soluble or insoluble in organic solvents. In acidic solutions and at elevated temperatures, degradation to smaller molecules occurs by cleavage of the glycosidic linkages. Selective cleavage of the glycosidic linkages can be achieved under carefully controlled reaction conditions.

Due to the presence of several asymmetric centres all aminoglycosides are optically active. The specific rotation $[\alpha]_D$ is dependent on the structural type[22a].

Structural type 2b $[\alpha]_D = 40–80°$
2c $[\alpha]_D = 120–160°$

Elucidation of the structures of newer aminoglycosides becomes easier due to the extensive practical experience developed in working with this class of compounds and because of the application of modern analytical methods.

The most difficult step is usually the careful purification, the separation of similar or isomeric minor components and the preparation of crystalline derivatives. Previously the formation of slightly soluble salts like helianthates and reineckates were used for the purification of the aminoglycosides. Today this method has been mostly replaced by chromatography with ion exchange resins. Cation exchange resins and adsorbents like silicagel, alumina and the dextran gels, are also very important for preparation and purification on a technical scale. For analytical determination of the aminoglycosides, paper chromatography, adsorption- and exchange-TLC, electrophoresis and gaschromatography are used[61–63]. A comprehensive survey is given by G. H. Wagman and M. J. Weinstein[64].

Microbiological assays have been developed for the quantitative determination of microgram amounts of the aminoglycosides in biological materials (blood, tissue). The serum concentration is directly related to the inhibition zone diameter after microbiological procedures[65–69]. Enzymatic methods for determining aminoglycoside antibiotic concentrations in serum are faster and more exact, namely acetyltransferases of R-factor carrying *E. coli* strains convert aminoglycosides in the presence of ^{14}C-acetyl-coenzyme A into a specifically acetylated radioactive derivative, which can be determined after separation on phosphocellulose paper[70–72].

An analogous method uses the enzyme AMP-transferase, which adenylates hydroxy groups of the aminoglycosides in the presence of ^{14}C-ATP or ^{3}H-ATP[73–75]. Radioimmunoassay techniques are also useful for the quantitative determination of aminoglycosides[76–79].

In structure-elucidation studies manifold use of physical methods such as IR- and NMR-spectroscopy[80–83], Reeves's Copper complexing method[84], and X-ray crystallographic analysis[85] are employed, supplemented by circular dichroism[86], ^{13}C-NMR (CMR)-spectroscopy[87–92], and mass spectroscopic studies[93–96]. Chemical work mainly serves the purpose of preparing suitable derivatives or definite fragments via methanolysis and mercaptanolysis, of the molecules.

In particular, the development of NMR-spectroscopy, e.g. INDOR-technique and CMR-spectroscopy, have facilitated the elucidation of structure and conformation of the carbohydrates. One of the main advantages of the CMR-technique is the large range (up to 200 ppm) over which the chemical shifts may be observed, thus making possible a better resolution than in PMR-analysis. The chemical shifts of C-atoms with similar environment in different glycosides are remarkably constant, permitting assignment for novel compounds through comparison of chemical shifts of known and unknown compounds. The measurement of off-resonance decoupled spectra provide additional useful data. The chemical shifts also indicate changes in conformation in the molecules.

Mass spectral determinations require that the compound under analysis should be sufficiently volatile and stable. For the aminoglycosides the volatility is increased by derivatisation e.g. N-acetyl-O-trimethylsilyl-, N,O-trimethylsilyl-, N-acetyl-N,O-methyl and N-salicyliden compounds.

To obtain a molecular ion, mild ionisation techniques are employed, i.e. electron impact (E. I.) with low ionization energy, chemical ionisation (C. I.) or field desorption methods (F.D.). The manner in which the various physical methods have been employed to elucidate the structure of the most important aminoglycoside antibiotics has been reviewed in detail by S. Umezawa[4].

III Characteristic Biological Properties and Methods for Their Improvement

Aminoglycosides are bactericidal antibiotics with a broad spectrum of antimicrobial activity[97b]. In particular, the pseudotrisaccharides of structural type 2c (Kanamycin B, DKB, Amikacin, Tobramycin, Gentamicin C-complex, Sisomicin and Verdamicin) having a deoxystreptamine moiety show high activity against gram-negative bacteria such as *Pseudomonas, Proteus, E. coli, Enterobacter, Serratia* and *Klebsiella,* which cause potentially fatal infections[97]. Rickettsiae, viruses, fungi, yeasts and protozoa are, however, resistant to aminoglycosides. Bacteriological in vitro data from different laboratories (MIC in mcg/ml) have been published[98]. Because of their strong dependence on the experimental parameters, the MIC values are only comparable if the experimental conditions are standardised. Some of the important parameters to be standardised are nutrient media of known composition with respect to alkali- and alkaline-earth ions, pH values and concentrations of serum[97a, 99–106].

A survey of MIC values and pharmacological data of aminoglycosides has been published by G. G. Grassi[7]. For the comparison of antibacterial activity of the different aminoglycosides, a cumulative graphical representation has often been used. The percentage of inhibition of clinically relevant strains is plotted against increasing antibiotic concentrations.

Figure 19 is an example of this kind of representation for tobramycin, gentamicin and amikacin. The order of potency of these compounds can vary depending on the number and sensitivity of the strains used.

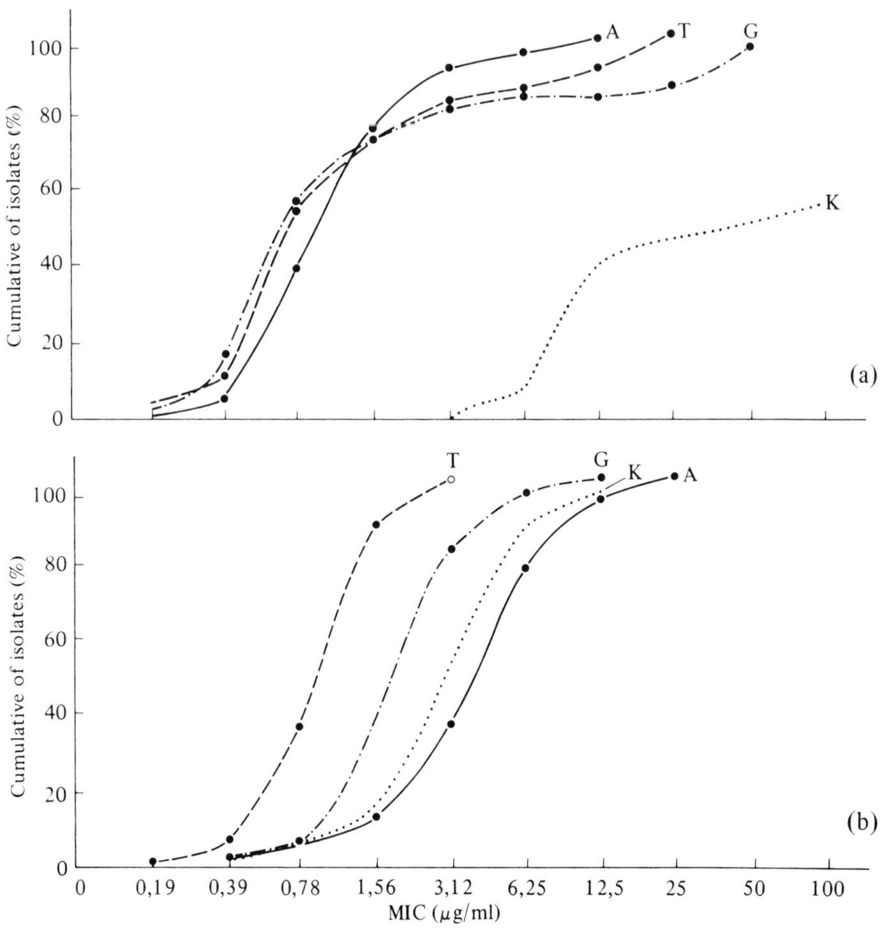

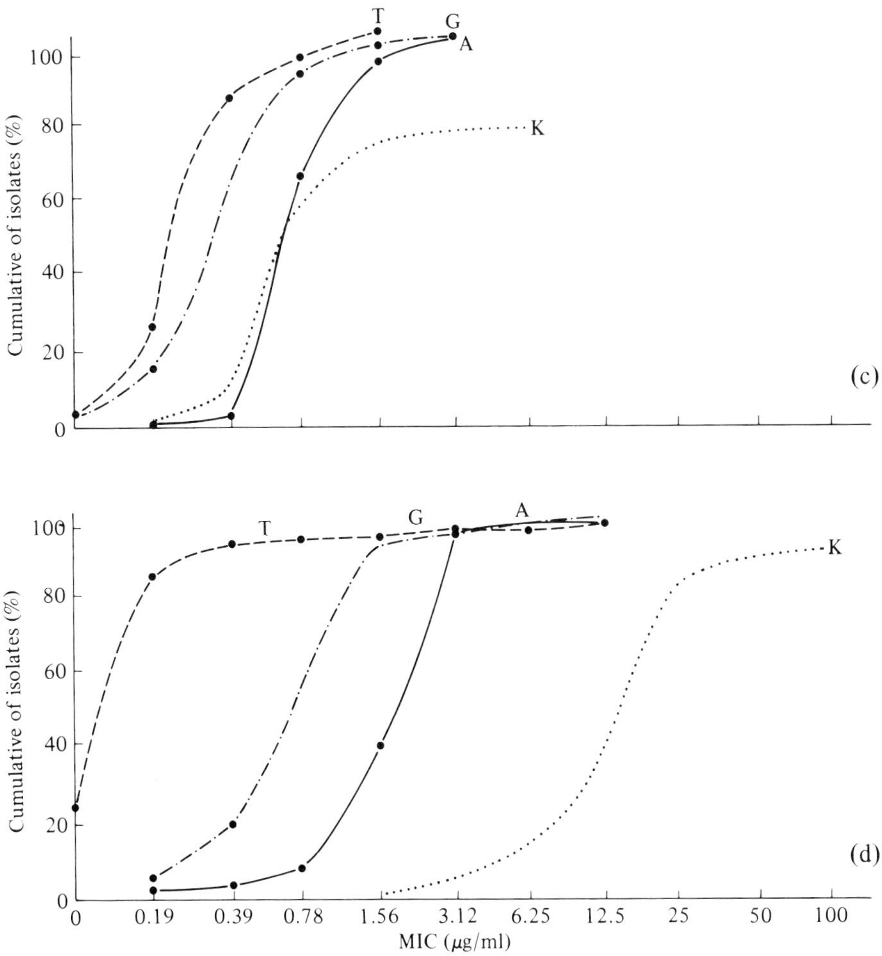

Fig. 19. **(a)** Serratia marcescens (50 strains); **(b)** Proteus mirabilis (56 strains); (c) Klebsiella (50 strains); **(d)** Pseudomonas aeruginosa (50 strains); Bodey, G. P., Stewart, D.: Antimicrob. Agents Chemoth. *4,* 186 (1973). A = Amikacin, G = Gentamicin; K = Kanamycin; T = Tobramycin

Tobramycin possesses a two- to fourfold higher activity than gentamicin and sisomicin against strains of *Pseudomonas aeruginosa* but a reverse order has been shown against *Serratia* and indole-negative *Proteus* species. The other gram-negative, pathogenic strains of bacteria have comparable sensitivity against these antibiotics[98c, 107)]. Several research groups have published additional biological data on other aminoglycosides: lividomycin[108)], butirosin[109)] and ribostamycin[110)]. Despite similarities in the structures of the most potent compounds, there are significant differences in sensitivity of strains belonging to the same bacterial species. Before aminoglycoside therapy is commenced, it is therefore necessary to test the sensitivity of the causative organisms against the different compounds[110)]. Combinations of penicillins or cephalosporins with aminoglycosides show synergistic effects[98a, 112)].

There are some reports of renal damage in cases of such therapeutic combinations, e.g. gentamicin and cephalothin[98a].

Streptomycin and dihydrostreptomycin are used only in special cases, e.g. in combination with isoniazid for the treatment of tuberculosis or in combination with tetracycline in the case of brucellosis[113].

In spite of the broad antibacterial spectrum of aminoglycosides and their effective utilisation in cases of severe bacterial diseases, world-wide attempts have been made to develop new aminoglycosides with improved properties and reduced side effects. A main goal in the development of new semi-synthetic products has been the selective protection and modification or elimination of the numerous hydroxyl- and amino groups, the exchange of specific aminosugars or aminocyclitols and the introduction of amino acids as in butirosin.

This work has contributed to an improvement in our knowledge of structure-activity relationships and mode of action.

What were the specific aims of these studies?

1. To enhance the inhibitory action (MIC value) in order to attack less sensitive microorganisms.
2. To overcome secondary resistance which has increased due to the frequent systemic and local administration of aminoglycosides.
3. To exclude or at least diminish side effects.
4. To attain enteral absorption (all aminoglycosides have to be administered by the parenteral route) and to optimise pharmacokinetic properties.

III.1 Total Syntheses

The successful clinical application of aminoglycoside antibiotics has stimulated worldwide intensive investigations into the possibilities of their synthesis. Due to the complicated stereochemistry and the numerous functional groups, it required many years before the first total synthesis of aminoglycosides was accomplished. Fundamental studies have been performed by S. Umezawa[4] and his research groups in Japan. The total synthesis of a natural product provides not only the final proof of its structure, but also shows new ways for the synthesis of structurally similar compounds, usually inaccessible by biosynthesis. Furthermore the provision of intermediates for antibacterial testing permits a clearer insight into the structure-activity relationships. All new, clinically important aminoglycoside antibiotics have been obtained by fermentation, with the exception of Amikacin and DKB (3′,4′-dideoxykanamycin B), which are semi-synthetic derivatives of kanamycin A and B.

The total synthesis of an aminoglycoside can be divided into the following steps:

1. Synthesis of monosaccharides or pseudodisaccharides, selective protection of functional groups and activation for the condensation reaction.
2. Formation of the α-glycosidic linkage and separation of possible isomeric by-products.
3. Deprotection, if necessary selectively in several steps, and eventually further condensation reactions.

Amino groups are often protected by ethoxycarbonyl-, benzyloxycarbonyl-, formyl-, acetyl- and 2,4-dinitrophenyl groups or by Schiff-bases. Azido groups are suitable precursors for amino groups. Hydroxyl groups can be protected by acetyl-, benzyl-, or trityl groups. Vicinal hydroxy groups form cyclic acetals (e.g. isopropylidene or cyclohexylidene). Vicinal OH, NH_2 groups may be converted to cyclic carbamates (oxazolidinones), which can be selectively cleaved. Numerous examples with bibliographical data are given in Section III.2.

One of the difficulties in the total synthesis of aminoglycosides lies in the formation of the glycosidic link. Stereochemical controls to receive definite regiospecific structures (especially the α-glycoside) are difficult and often result in poor yields. A comprehensive survey on problems of O-glycoside synthesis has been published by G. Wulff and G. Röhle[114]. Since the successful studies of Koenigs and Knorr[115], 1-halosugars have been mainly used in different solvents in the presence of acid and water scavengers (Ag_2O, Ag_2CO_3, $AgClO_4$, $Hg(CN)_2$, $HgBr_2$ and drierite). These methods are not stereospecific and they result in a mixture of anomers. Lemieux et al.[116] found an efficient glycoside synthesis, which utilizes the cis-directing effect of a C-2-nitroso group in a 1-halosugar. Dimeric nitroso sugars of type (I), obtained through cis-addition of nitrosylchloride on acetylated glycals, condense even with complicated sugar derivatives (ROH) via the intermediate (II) with high stereoselectivity to compound (III) (Fig. 20). Following this procedure the proportion of trans glycoside formation is small. The cleavage of oximino compound (III) can be achieved with acetaldehyde/HCl or titanium chloride. The keto function is reducible with $NaBH_4$ or B_2H_6.

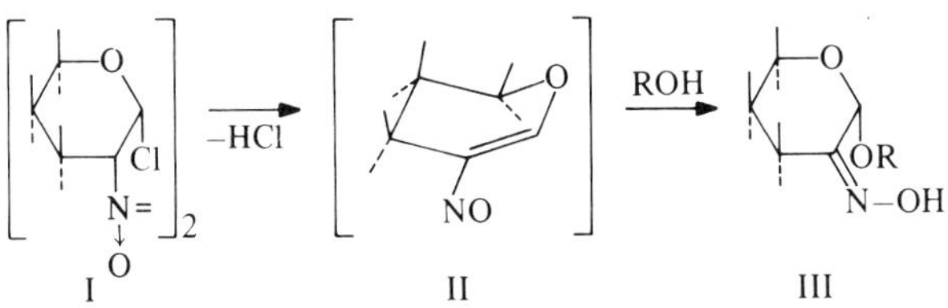

Fig. 20.

III.1.1 Total Synthesis of Aminoglycosides with Two Carbohydrate Units and an α-D-Glycosidic Linkage

Several compounds of this type result directly either by fermentation (e.g. neomycin A = neamine) or by the acidic cleavage of higher condensed aminoglycosides. Some of them, such as neamine and paroamine, are weakly effective and not useful for practical application.

Basic studies concerning their total synthesis have been undertaken by S. Umezawa and co-workers[4]. All the 0-(amino-deoxy-α-D-glucopyranosyl)-2-deoxystreptamines listed in Table 2 have been synthesized by a modified Koenigs-Knorr condensation or via the nitrosyl chloride method mentioned above.

The total synthesis of kasugamycin, a two-unit type aminoglycoside, produced by *Streptomyces kasugalusis,* has also been described.It is a potent inhibitor of *Pericularia oryzae,* a plant pathogen[117].

Table 2.

Name	Structure	Ref.
69 4-O-(4-amino-4-deoxy-α-D-glycopyranosyl)-2-deoxystreptamine	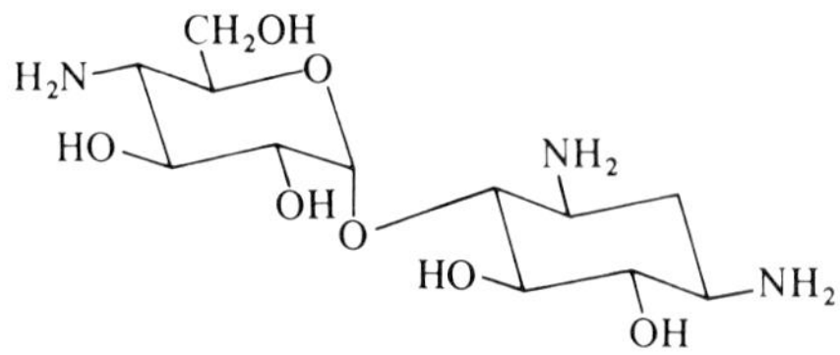	118) No antibact. act.
70 6-O-(4-amino-4-deoxy-α-D-glycopyranosyl)-2-deoxystreptamine	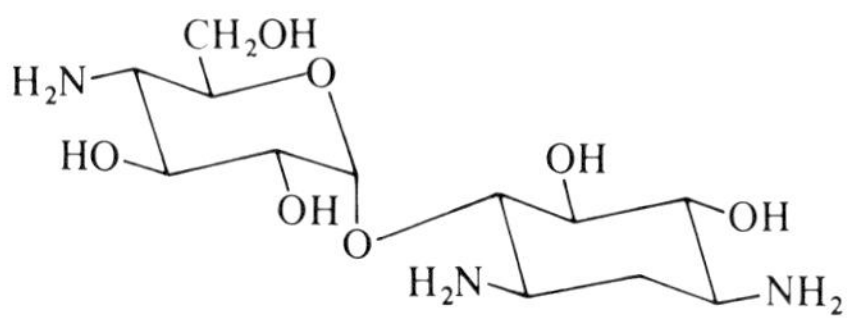	118) No antibact. act.
71 5-O-(2-amino-2-deoxy-α-D-glycopyranosyl)-2-deoxystreptamine	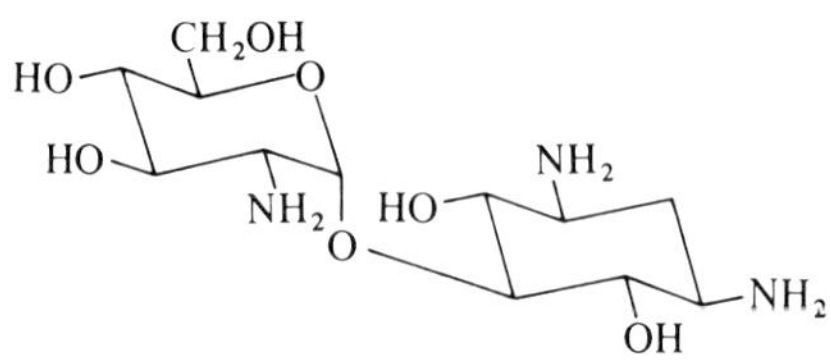	4) p. 148, 149
72 4-O-α-D-glucopyranosyl-2-deoxystreptamine	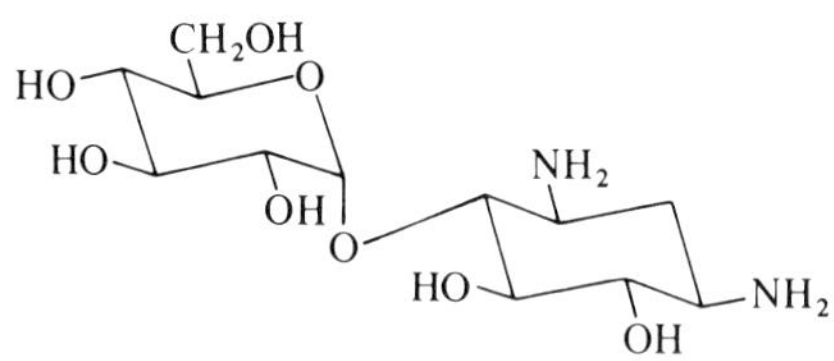	118, 119)
73 6-O-α-D-glucopyranosyl-2-deoxystreptamine	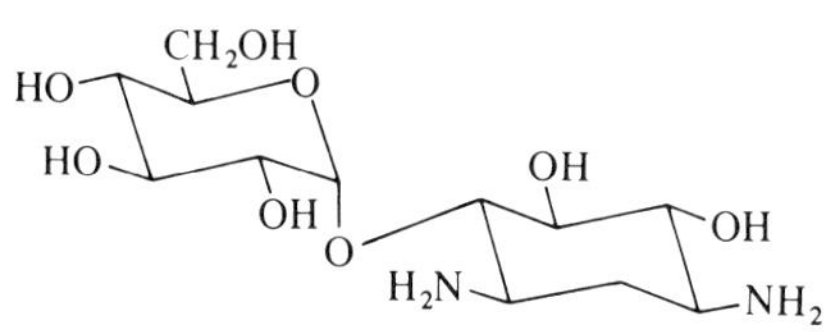	118, 119)
74 5-O-α-D-glycopyranosyl-2-deoxystreptamine	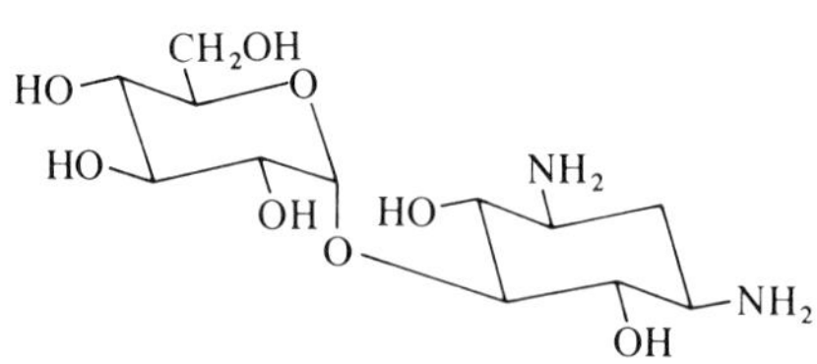	119)
75 Neamine (neomycin A)	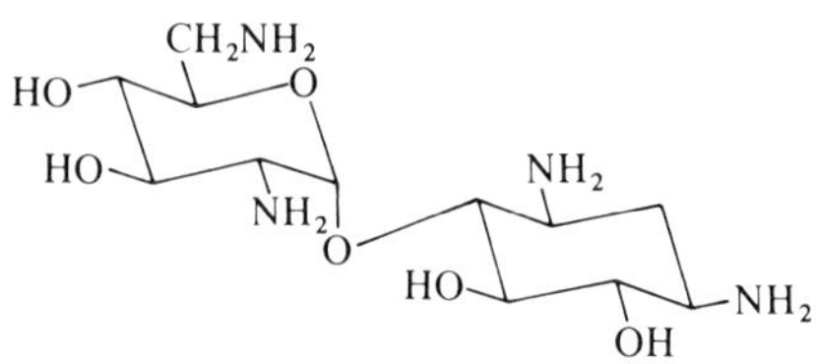	120)

Table 2 (continued)

Name	Structure	Ref.
76 6-O-(3,6-diamino-3,6 dideoxy-α-D-glucopyranosyl)-2-deoxy-streptamine	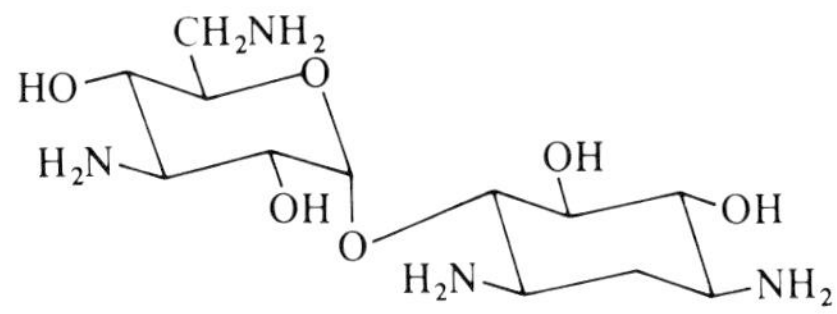	121) No antibact. act.
77 6-O-(3-amino-3-deoxy--D-glucopyranosyl)-2-deoxy-streptamine	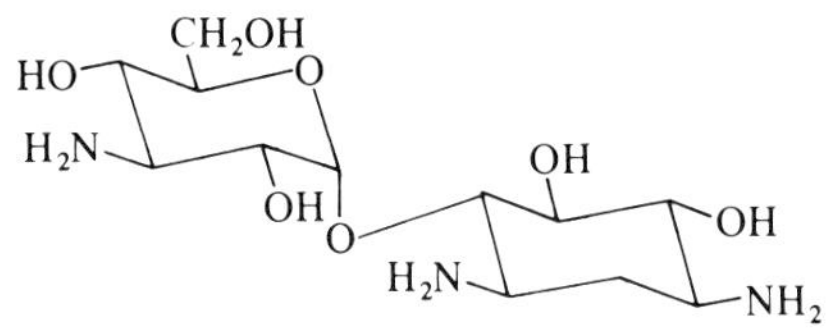	122, 123)
78 Paromamine	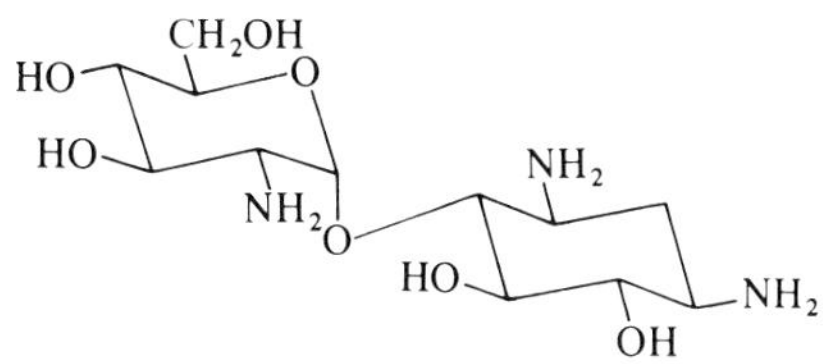	No antibact. act. 124, 125)
79 6-O-(2-amino-2-deoxy--D-glucopyranosyl)-2-deoxystreptamine	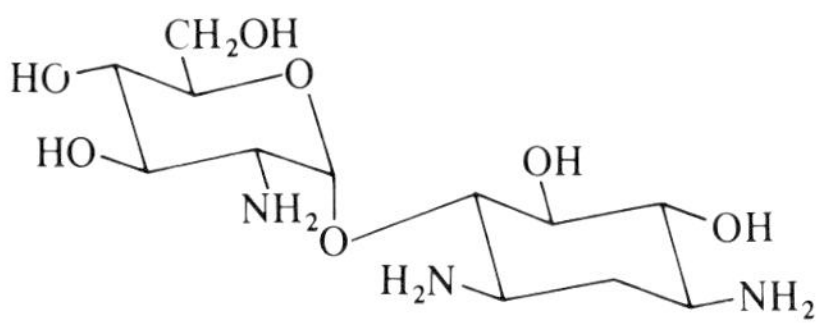	125)
80 4-O-(6-amino-6-deoxy--D-glucopyranosyl)-2-deoxystreptamine and tri-N-acetyl derivative	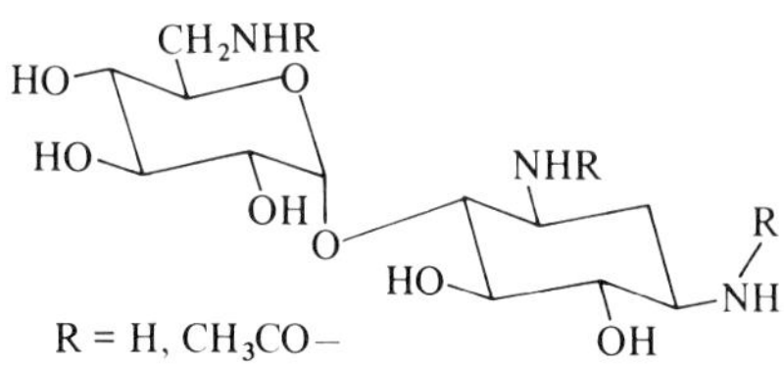 R = H, CH₃CO–	126) R=H → antibact. act.
81 6-O-(6-amino-6-deoxy--D-glucopyranosyl)-2-deoxystreptamine and tri-N-acetyl derivative	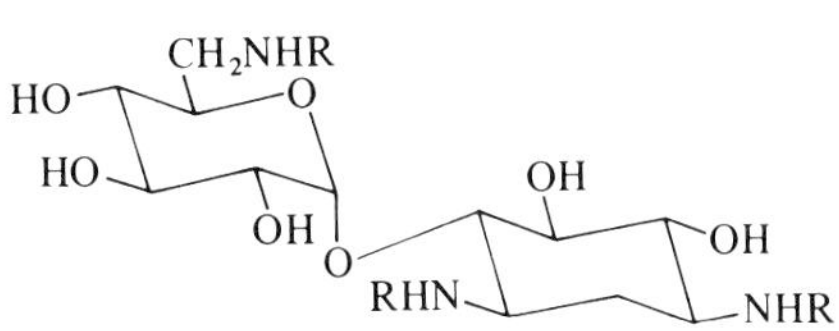	126)

III.1.2 Total Synthesis of Aminoglycosides with Three Carbohydrate Units

Kanamycin A, the oldest compound of this type, has been synthesised by two Japanese research groups. S. Umezawa and co-workers[127] linked the protected 6-0-(3-amino-3-deoxy-α-D-glucopyranosyl)-2-deoxystreptamine (Table 2, *77*) with the blocked glycosylchloride, prepared from 6-amino-6-deoxyglucose. Nakajiama and co-workers[128] have reported the synthesis of kanamycin A from the protected 4-0-(6-amino-6-deoxy-α-D-glucopyranosyl)-2-deoxystreptamine (Table 2, *80*) and the tri-0-benzylglycosylchloride, obtained from 3-acetamido-3-deoxy-glucose.

Kanamycin B has also been synthesized by S. Umezawa et al.[129] by coupling a suitably protected neamine (Table 2, *75*) with 3-amino-3-deoxy-D-glucose. The synthesis of kanamycin C has been accomplished by the same researchers[130] through the glycosylation of 4′, 6′-isopropylidine-1,3,2′-tri-N-benzyloxycarbonylparomamine (Table 2, *78*) with tri-0-benzylglycosylchoride, obtained from 3-acetamido-3-deoxyglucose.

Tobramycin (3′-deoxykanamycin B) has been prepared from kanamycin B, starting with the protected 3′-tosyl derivative[131] (Sect. III.2).

Ito and co-workers[132] have reported the synthesis of ribostamycin from 3′,4′-dibenzyl-tetra-N-(benzyloxycarbonyl)-neamine and 2,3,5-tri-0-benzyl-D-ribofuranosylchloride (Sect. III.2.). An alternative synthesis of ribostamycin and biologically inactive isomers starts with 4-acetyl-N,N′-dicarbo-benzoxy-2-deoxystreptamine, which reacts with 2,3,4-tri-0-benzoyl-D-ribofuranosyl chloride in the presence of $AgClO_4$-Ag_2CO_3 to the tricyclic 5-0-β-D-ribofuranosyl-2-deoxystreptamine[133]. Butirosin has been synthesised by S. Umezawa and co-workers[134] from protected ribostamycin (Sect. III.2.3).

The glycosylation of the appropriately protected neamine, 3′-deoxyneamine and 3′,4′-dideoxyneamine with the glycosyl chloride accessible from desosamine leads to biologically interesting aminotrisaccharides[135, 136]. Some further aminoglycosides related to kanamycin have been synthesised (Fig. 21):

a. NK 1001 (I), an antibiotic, formed by mutants of *Streptomyces kanamyceticus*[137].
b. 4,6-di-0-(6-amino-6-deoxy-α-D-glucopyranosyl)-2-deoxystreptamine (II)[137].
c. 4,6-di-0-α-D-glucopyranosyl-2-deoxystreptamine (III)[138].

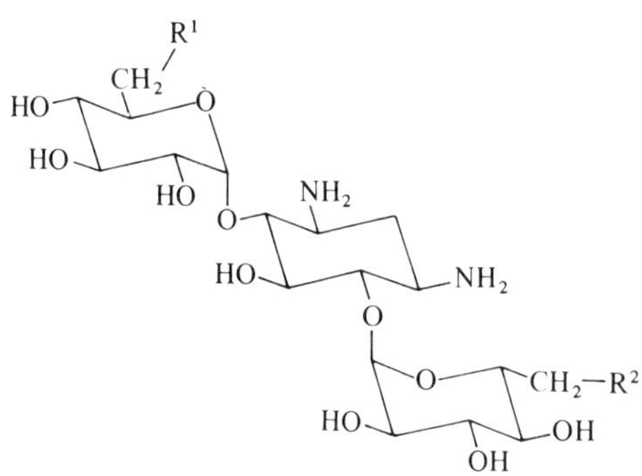

	R^1	R^2
I	NH_2	OH
II	NH_2	NH_2
III	OH	OH

Fig. 21.

Recently, a total synthesis of neomycin C, starting with ribostamycin, was reported[139].

The most difficult total syntheses in this class of pseudotrisaccharides were those of streptomycin and dihydrostreptomycin. About 35 years after their discovery and the elucidation of their structure, S. Umezawa and colleagues[140)] succeeded in synthesising these compounds. The glycosyl derivative (I), prepared through several steps from streptobiosaminide, was condensed with the protected streptidine (II), to yield dihydrostreptomycin, after deprotection. Subsequent oxidation with DMSO/dicyclohexylcarbodiimide (DCC) produced streptomycin[140a)].

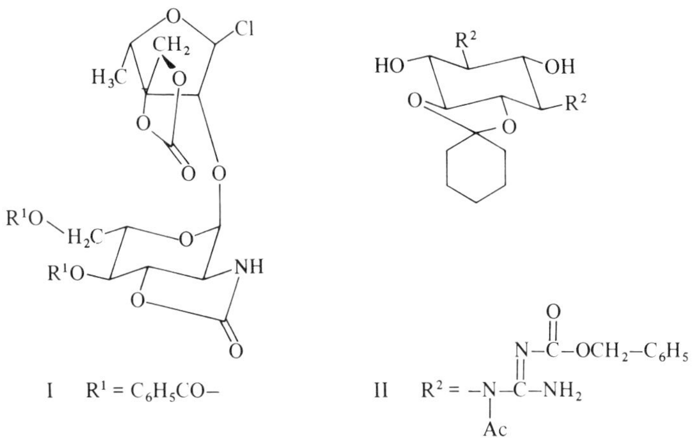

Fig. 22.

Streptomycin and dihydrostreptomycin are inactivated by R-factor-carrying *E.coli* or *Pseudomonas* strains through O-phosphorylation or O-adenylation in the 3″-position. 3″-Deoxystreptomycin is the first modified streptomycin in which this type of resistance does not develop[141)].

Thus, the treatment of D-N-acetyl-di-N-benzyloxycarbonyl-4,5-0-cyclohexylidine-streptidine with the glycosyl chloride from 2-0-(2-acetamido-4,6-di-0-acetyl-2-dideoxy-N-methyl-L-glucopyranosyl)-3,3′-0-carbonyl-dihydrostreptose (17 steps) in dichlormethane, containing mercuric cyanide and a molecular sieve, gave after chromatography and deprotection 3″-deoxydihydrostreptomycin.

III.2 Chemical Modifications of Aminoglycosides

The aminoglycosides readily accessible by fermentation or by degradation of natural products are the best starting materials for partial synthesis. The protection and/or the activation of the numerous amino- and hydroxyl functions, which show only little differences in reactivity, cause many problems for controlled chemical modification. The amino groups of gentamicin and neamine, for example, possess similar basicity due to their chemical environment. After alkylation- and acylation reactions, mixtures of mono- or poly-substituted products very often arise, which are often difficult to separate. Therefore, the right use of specific reagents, protecting groups and reaction conditions is a basic requirement for controlled partial synthesis.

Vicinal hydroxyl groups can be protected advantageously by conversion into cyclohexylidene derivatives according to Bissett[142]. This reaction has been used by Umezawa[143] for the differentiation of the hydroxyl groups in neamine and for the synthesis of 3',4'- and 5,6-modified neamine derivatives:

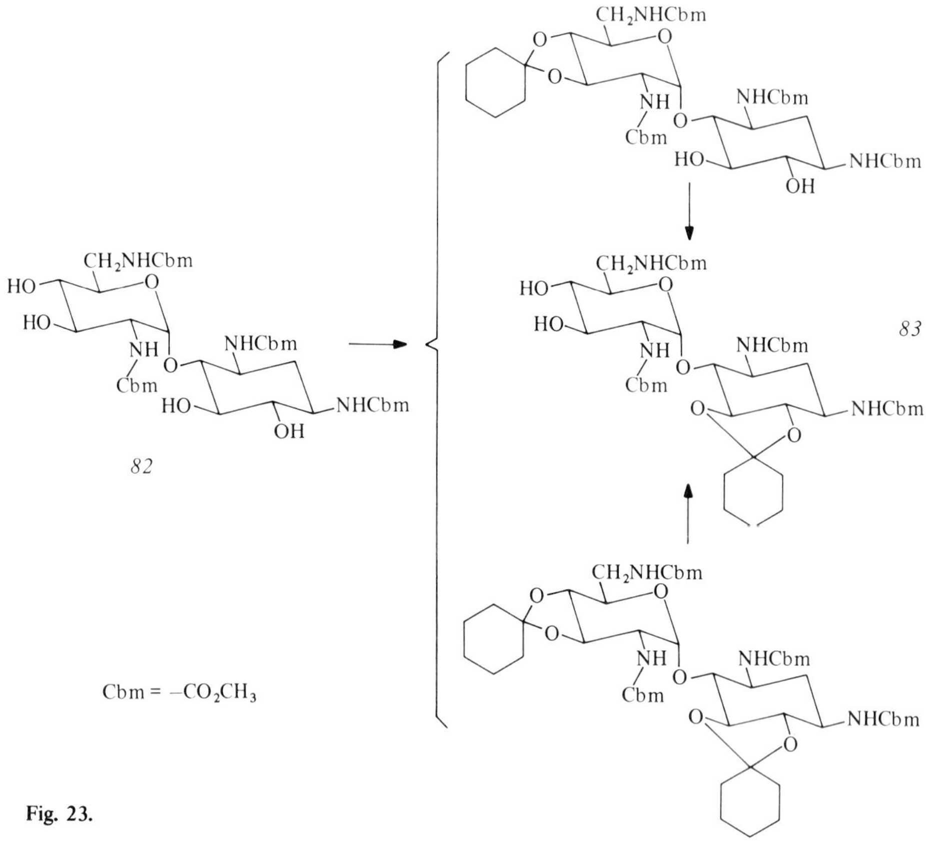

Fig. 23.

1,2-aminoalcohol structures have been protected through cyclisation to oxazolidinones using N,N'-carbonyldiimidazole according to K. Miyai and P. H. Gross[144].

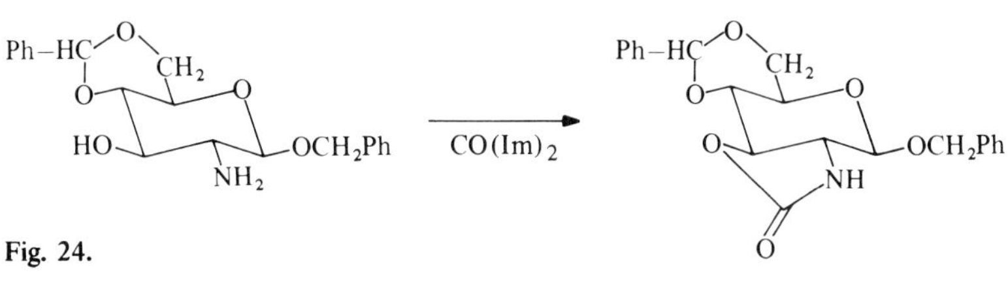

Fig. 24.

As shown by S. Umezawa[145], this reaction can be conducted in water-containing solvents, so that only the amino alcohol function reacts, whereas the other hydroxyl groups are not altered.

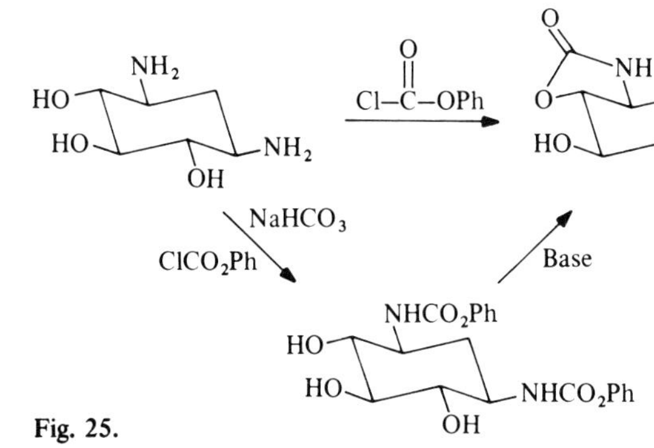

Fig. 25.

The discovery that benzyloxycarbonyl protected amino groups with adjacent hydroxyl functions cyclise in the presence of NaH has led to a broad application of the technique[146]. The oxazolidinone ring can be selectively cleaved with barium hydroxide to give the free amino alcohol group:

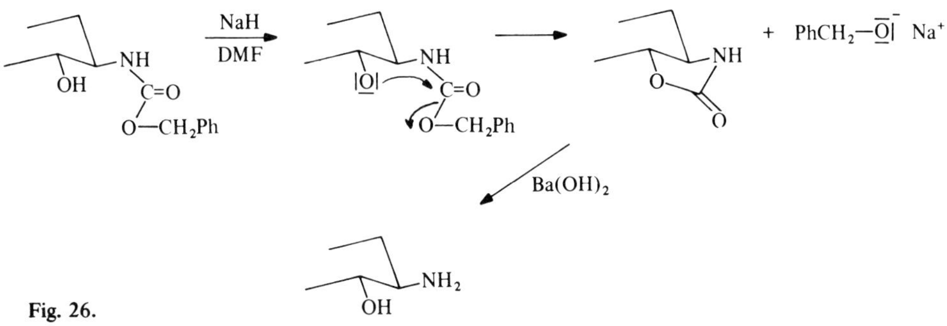

Fig. 26.

This reaction is of importance for the 1-N-acylation of type 2a,b-aminoglycosides.

An alternative method for the selective acylation of the 1-amino group has been described in a patent for xylostasin, ribostamycin and kanamycin A and B[147]. These aminoglycosides, performylated with p-nitrophenylformate, react within a few days with aqueous ammonia to provide by selective cleavage the free 1-amino groups, which can then be selectively acylated. The synthesis of numerous deoxy derivatives can be achieved following the procedure of A. Cohen and R. S. Tipson[148, 149]: the vicinal di-0-mesyl and di-0-tosyl derivatives have been converted to the appropriate alkenes with sodium iodide/zinc dust in dimethylformamide, which can then be hydrogenated. The removal of vicinal mesyloxy groups using naphthalene-sodium to afford the eno compound has also been described[150]. The following scheme shows further routes for the preparation of monodeoxy- and aminodeoxy derivatives:

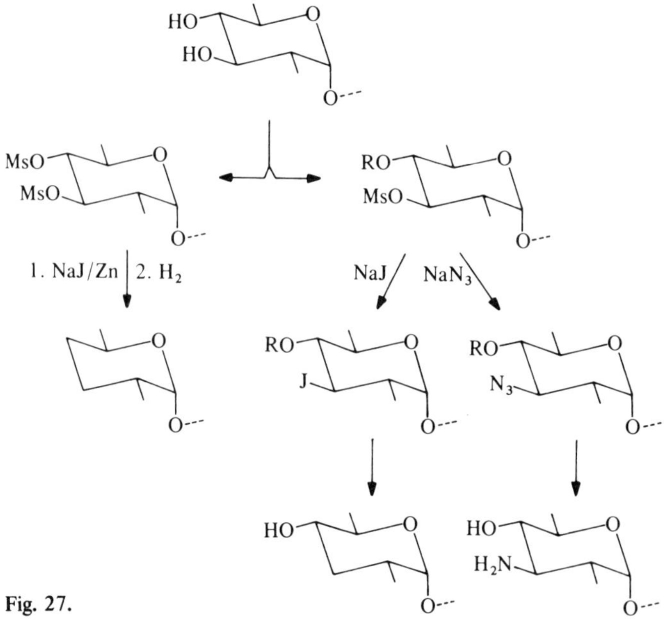

Fig. 27.

III.2.1 Streptidine Containing Aminoglycosides

A great number of attempts have been made to improve the pharmacological and antibacterial properties of streptomycin. Most of the synthetic work dates from the years 1960–1968. Since then only a few new streptomycin derivatives have been described[151].

The most important variations (Fig. 28) are: redox (1)- and condensation reactions (2) with the aldehyde group, degradation- and synthetic reactions with the guanidine groups (3) and reactions with the methylamino group of L-hexosamine (4)[152].

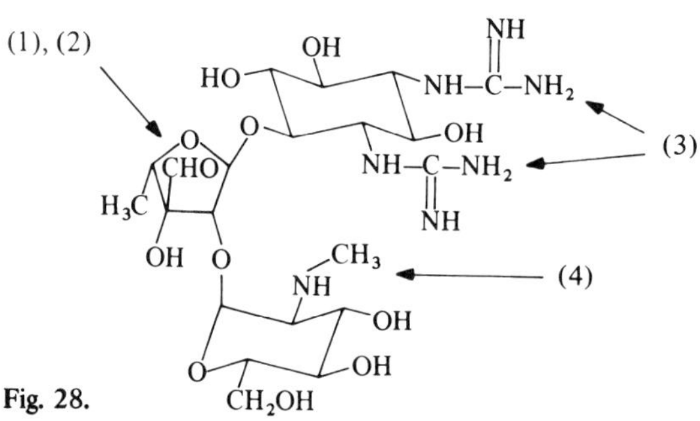

Fig. 28.

From all the streptomycin derivatives, only the dihydrostreptomycin 6, which technically has been prepared by fermentation, is of importance. Bluensomycin plays no important role in therapy.

III.2.2 4-Substituted Deoxystreptamine Antibiotics – Neamine

Neamine 8, a common structural element in aminoglycoside antibiotics, can be obtained easily through acidic hydrolysis of neomycin B. The modification of its functional groups (Table 3) has a great influence on chemotherapeutic properties.

Table 3.

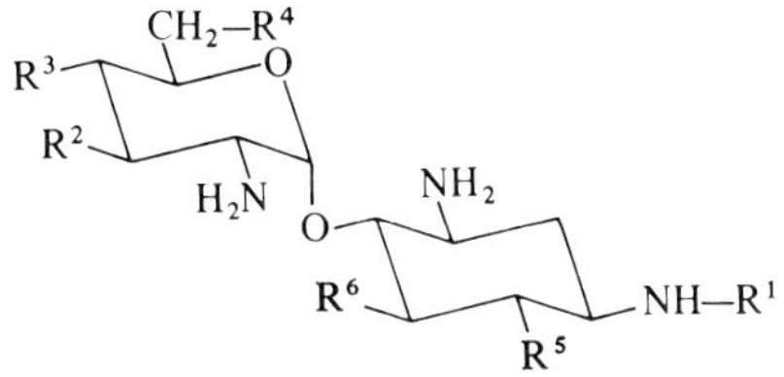

	R^1	R^2	R^3	R^4	R^5	R^6	Ref.
86 3'-Deoxyneamine	H	H	OH	NH_2	OH	OH	153)
87 3',4'-Dideoxyneamine	H	H	H	NH_2	OH	OH	143)
88 3'-Epi-4'-deoxyneamine	H	OH^a	H	NH_2	OH	OH	155)
89 3',4',5,6-Tetradeoxyneamine	H	H	H	NH_2	H	H	158, 158c)
90 5,6-Dideoxyneamine	H	OH	OH	NH_2	H	H	158c)
91 3'-Deoxy-6'-methylneamine	H	H	OH	$NHCH_3$	OH	OH	153)
92 3'-O-Methylneamine	H	OCH_3	OH	NH_2	OH	OH	143)
93 4'-O-Methylneamine	H	OH	OCH_3	NH_2	OH	OH	143)
94 3'-Mercaptoneamine	H	SH	OH	NH_2	OH	OH	157)
95 4'-Mercaptoneamine	H	OH	SH	NH_2	OH	OH	157)
96 Acylneamine (HABA and Analogue)	ACYL	OH	OH	NH_2	OH	OH	161, 162, 164, 166)
97 Acyl-3'-deoxyneamine	ACYL	H	OH	NH_2	OH	OH	166)
98 Acyl-3',4'-dideoxyneamine	ACYL	H	H	NH_2	OH	OH	164–166)

a epimer.

The synthesis of 3'-deoxyneamine 86[153)] and 3',4'-di-deoxyneamine 87[143)] is shown in the following scheme:

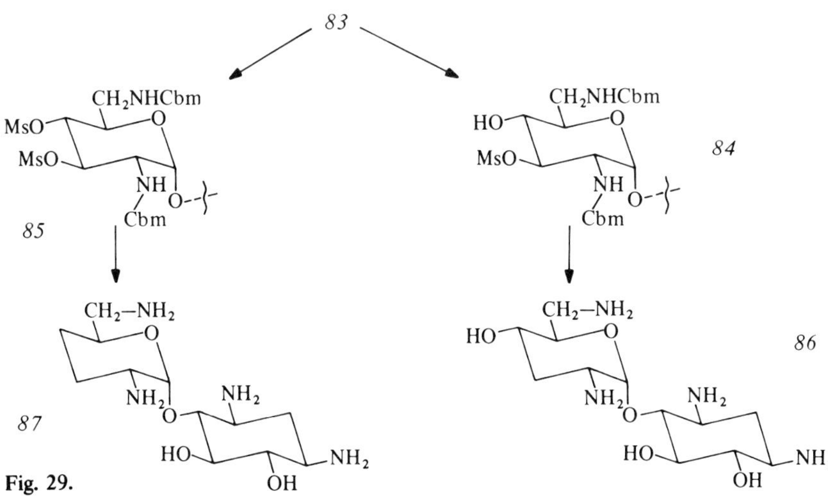

Fig. 29.

The mesylation of *83* to *85* followed by iodination and catalytic hydrogenolysis gives *87*. Compound *86* has been synthesised via controlled mesylation or tosylation at the 3′-position, iodination and reduction. A patent from Meiji[154] describes the conversion of the 3′-tosyl derivative into the 3′,4′-epoxy compound and the catalytic reduction to *86*. 3′-Epi-4′-deoxyneamine, *88*, is prepared using $NaBH_4$ for the reduction[155]. The preparation of 3′,4′-epoxyneamine derivatives *100, 101*[154–156] and 3′- and 4′-mercaptoneamines *94, 95*[157] is shown in the following figure:

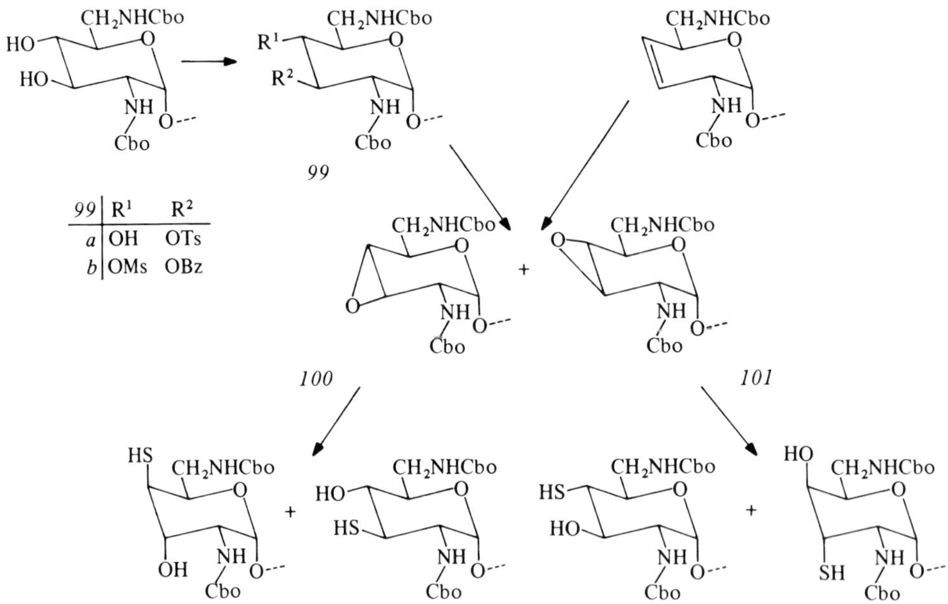

99	R¹	R²
a	OH	OTs
b	OMs	OBz

Fig. 30.

3′4′,5,6-Tetradeoxyneamine *89*[158] and 5,6-dideoxyneamine *90*[158c] have been synthesised from the N-protected and mesylated neamine through reaction with sodium iodide/zinc and subsequent hydrogenolysis.

3′-Methoxy-*92* and 4′-methoxyneamine *93*[142] are less active than neamine and inactive against resistant strains of bacteria. The 3′-deoxy derivatives show antibacterial activity against such resistant bacteria, which are known to inactivate aminoglycosides by O-phosphorylation at the 3′-position.

The observation that butirosin *21* is more effective than ribostamycin *18*, provided the impetus to introduce the (S)-2-hydroxy-4-aminobutyric acid substituent into other aminoglycosides. Starting from 1-N-HABA neamine[159, 160], obtained by acidic hydrolysis of butirosin, several hydroxyaminoacyl groups of different configurations have been introduced[161] by the following steps: Protection of the free

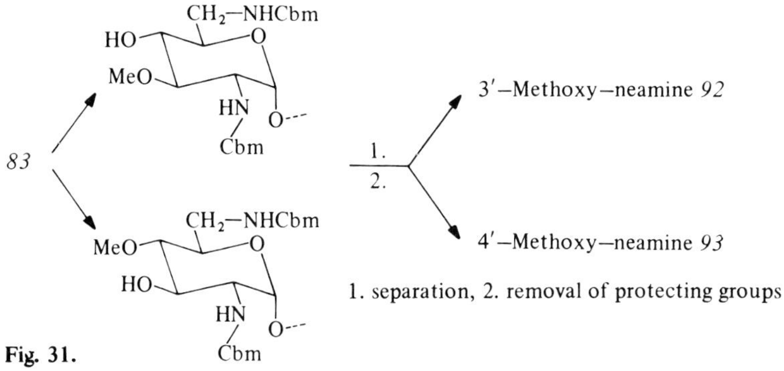

Fig. 31.

amino functions by the phenylsulfonyl group, removal of the HABA moiety under basic conditions and acylation of the liberated 1-amino function, followed by the removal of the protecting groups.

E. Akita[162)] acylated neamine trifluoroacetate using the N-protected S(-)-2-hydroxy-4-aminobutyric acid (HABA) in the presence of DCC.

The O-acylated derivative underwent a O → N acyl migration from the 6–0 to the 1-amino group by treatment with hydrazine. Similarly the 3',4'-dideoxy compound has been prepared[163)]. More often Umezawa's method[146)] was used for the controlled 1-N-acylation. Compound *102a* which is easily obtained by cleavage of the oxazolidinone ring with $Ba(OH)_2$ can be modified at the 1-amino function[164–166)].

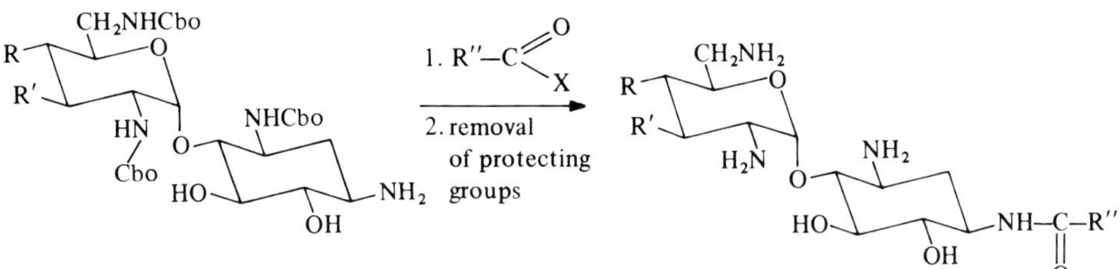

	R	R'
101a	H	H
b	–O– (cyclohexylidene)	–O–

Fig. 32.

III.2.3 4,5-Disubstituted Deoxystreptamine Antibiotics

III.2.3.1 Ribostamycins, Xylostasins, Butirosins. Numerous semi-synthetic derivatives of ribostamycin, xylostasin and butirosin A and B have been prepared. Table 4 represents a survey of these derivatives.

Table 4

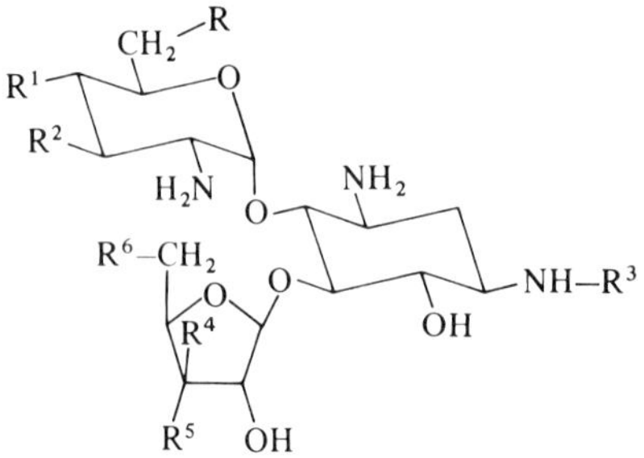

		R	R^1	R^2	R^3	R^4	R^5	R^6	Ref.
104	3'-Deoxyribostamycin	NH_2	OH	H	H	H	OH	OH	169)
105	3',4'-Dideoxyribostamycin	NH_2	H	H	H	H	OH	OH	171)
106	3',4',5''-Trideoxyribostamycin	NH_2	H	H	H	H	OH	H	171)
107	3',4'-Dideoxyxylostasin	NH_2	H	H	H	OH	H	OH	170)
108	1-N-acyl-3',4'-dideoxyribostamycin	NH_2	H	H	ACYL*	H	OH	OH	166)
109	1-N-acylribostamycin (-xylostasin)	NH_2	OH	OH	ACYL*	H (OH)	OH (H)	OH	147, 159) 164, 167)
110	6'-N-acylribostamycin (-xylostasin)	NH-ACYL*	OH	OH	H	H (OH)	OH (H)	OH	159)
111	3'-Deoxybutirosin B	NH_2	OH	H	HABA	H	OH	OH	168)
112	3',4'-Dideoxybutirosin A	NH_2	H	H	HABA	OH	H	OH	173)
113	3',4'-Dideoxybutirosin B	NH_2	H	H	HABA	H	OH	OH	174)
114	5''-Amino-5''-deoxybutirosin A	NH_2	OH	OH	HABA	OH	H	NH_2	176)
115	5''-Amino-5''-deoxybutirosin B	NH_2	OH	OH	HABA	H	OH	OH	176)
116	5''-Amino-3',5''-dideoxybutirosin A	NH_2	OH	H	HABA	OH	H	NH_2	175)
117	5''-Amino-4',5''-dideoxybutirosin A	NH_2	H	OH	HABA	OH	H	NH_2	177)
118	5''-Amino-3',4',5''-trideoxybutirosin A	NH_2	H	H	HABA	OH	H	NH_2	178, 179)
119	Deneosaminyl-lividomycin	OH	OH	H	H	H	OH	OH	180)

ACYL* = HABA and analogues

To synthesise butirosin and its analogues, 6′-N-protected ribostamycin was acylated via the active ester with the following acids R–OH or R′–OH[159]:

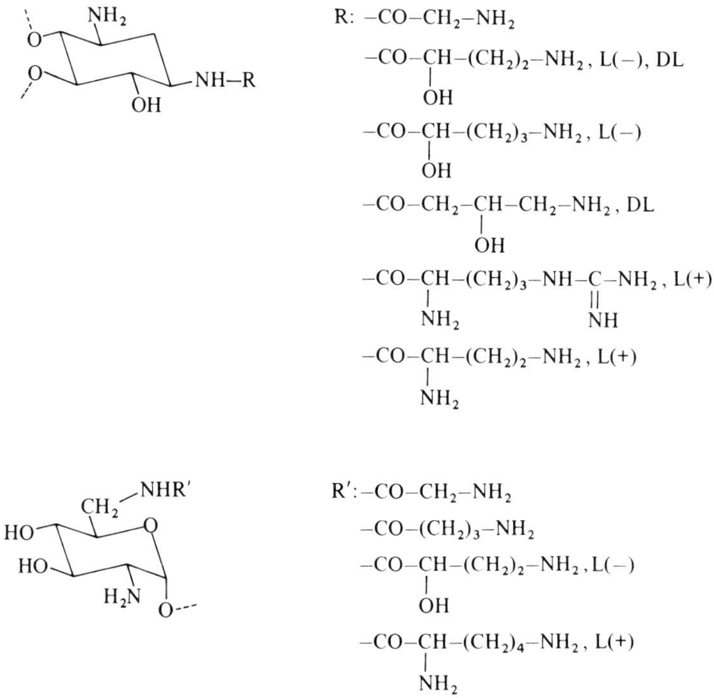

Fig. 33.

6′-N-acylated compounds with low antibacterial activity were among the products formed when ribostamycin was acylated. A comprehensive study of the influence of HABA and its analogues on the antibacterial activity of the aminoglycosides has been published by T. H. Haskell[167]. A large number of 1-N-acylated butirosin derivatives was synthesised in the following way (Fig. 34):

Butirosin is converted to the N-protected derivative *120* by reaction with dimedone. Treatment with $Ba(OH)_2$ gives a protected ribostamycin with a free 1-amino group *121*, from which 1-N-butirosin derivatives *109* can be obtained by acylation with various amino acids, followed by removal of the protecting groups. The derivatives with 2-hydroxy-ω-aminoacyl moieties containing two to three C-atoms show the best activities. The change of the asymmetric centres of the amino acids from the S- to the R-configuration causes a decrease in activity. (For further details regarding the structure-activity relationship, see Chap. IV).

Another possibility for the direct synthesis of 1-N-acylated ribostamycins (i.e. butirosin derivatives) via the 3,2′,6′-triformylribostamycin has been described in Chap. III.2[147].

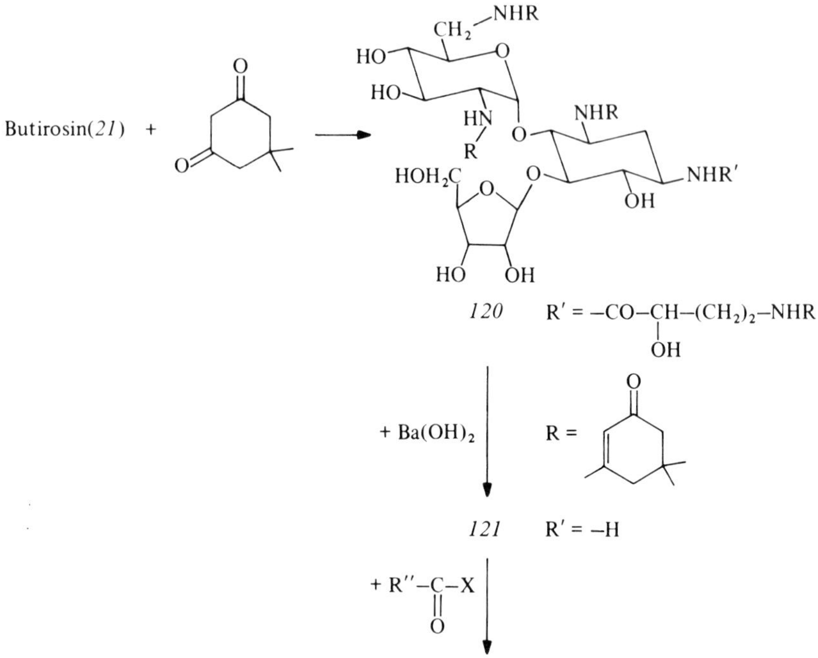

Fig. 34.

The previously mentioned method of Umezawa et al.[146] can be used successfully for the synthesis of 3′-deoxy- and 3′,4′-dideoxybutirosin A and B from ribostamycin or xylostasin respectively.

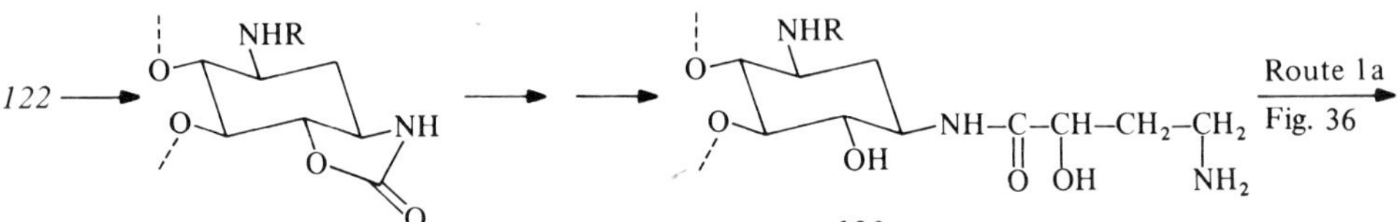

Fig. 35.

Starting with the protected ribostamycin *122* (Fig. 36), into which the HABA-moiety can be introduced via the cyclic carbamate, (Fig. 36) 3′,4′-dideoxybutirosin B *113* can be prepared following a route analogous to 1a and 3′-deoxybutirosin B *111* can be synthesised by a route similar to 1b[168].

The synthesis of the mono-, di- and trideoxy derivatives of butirosin and ribostamycin via dehydroxylation is summarised in Fig. 36.

3′-Deoxyribostamycin *104*[169], 3′,4′-dideoxyxylostasin[170], 3′,4′-dideoxyribostamycin *105* and 3′,4′,5″-trideoxyribostamycin *106*[171] as well as 3′-deoxybutirosin A[172], 3′-deoxybutirosin B *111*[168], 3′,4′-dideoxybutirosin A[173], 3′,4′-dideoxybutirosin B *113*[174] and 5″-amino-3′,5″-dideoxybutirosin A[175] have been synthesised in a similar manner. The course of the reaction is the same for butirosin A

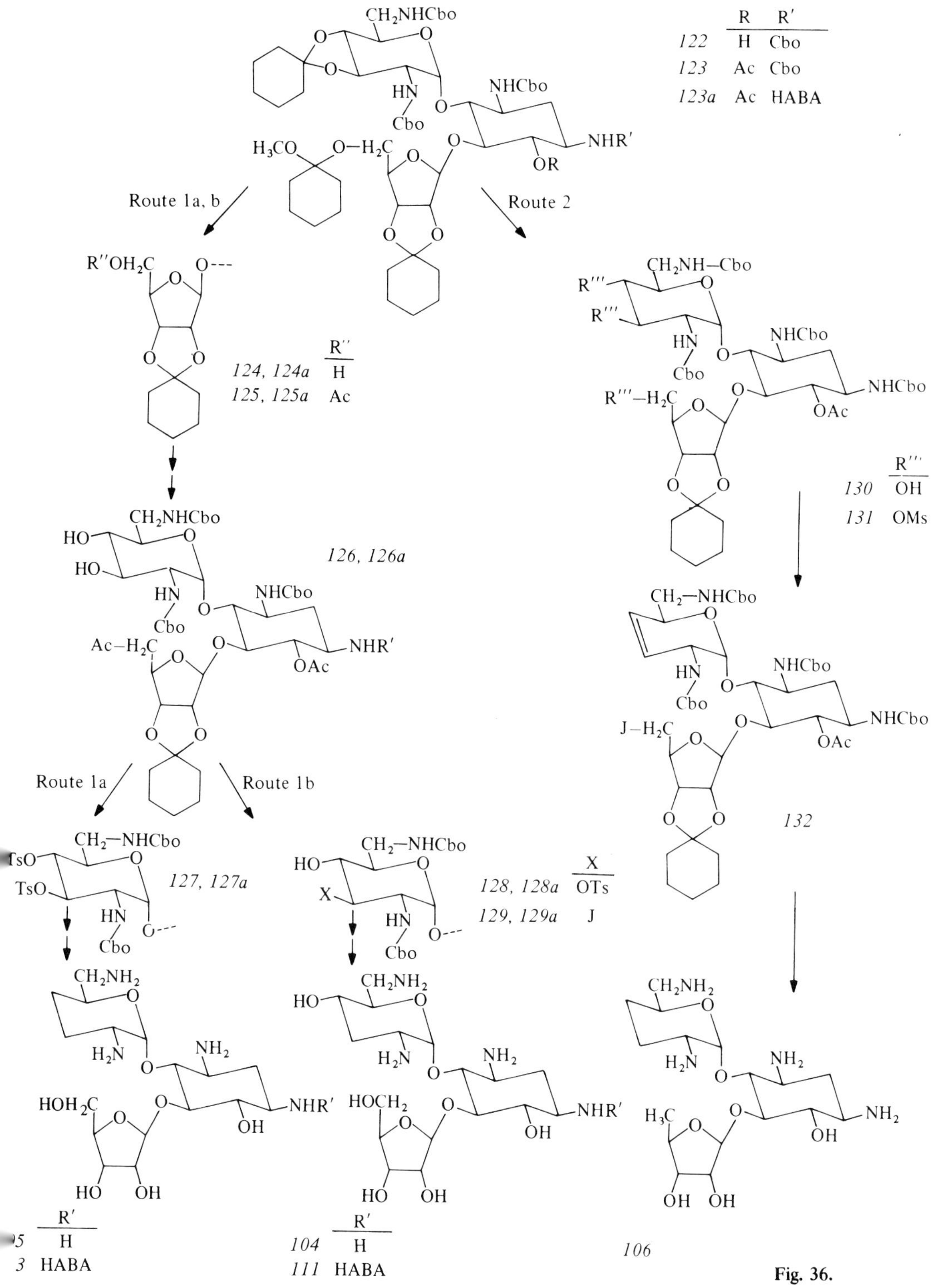
R R′
122 H Cbo
123 Ac Cbo
123a Ac HABA
Route 1a, b
Route 2
R″
124, 124a H
125, 125a Ac
126, 126a
R‴
130 OH
131 OMs
132
Route 1a
Route 1b
127, 127a
X
128, 128a OTs
129, 129a J
R′
95 H
3 HABA
R′
104 H
111 HABA
106

Fig. 36.

and B or ribostamycin and xylostasin, although the intermediates are different due to the different chirality at the C-3″ position (Fig. 37).

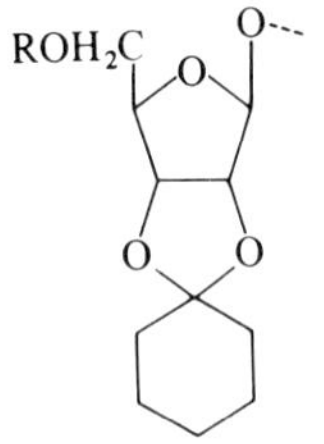

Butirosin B

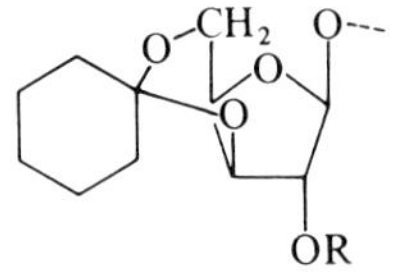

Butirosin A

Fig. 37.

Since the removal of the 5″-hydroxyl group in ribostamycin led to a decrease of activity, the 5″-hydroxyl group in butirosin A and B[176] and in 4′-deoxybutirosin A[177] was substituted with an amino group. In these products the activity was enhanced.

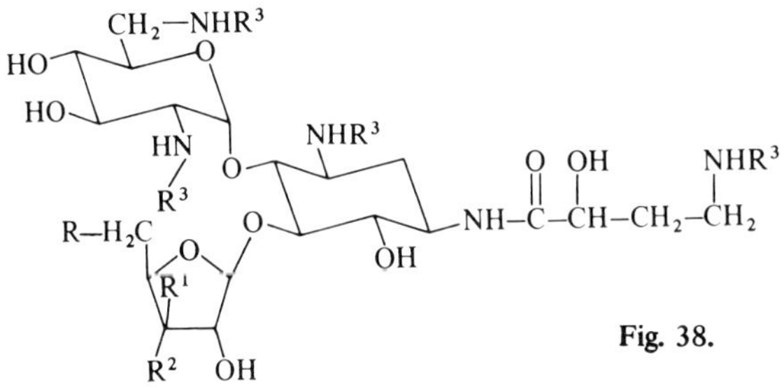

Fig. 38.

	R	R^1	R^2	R^3
133	OH	OH	H	F_3C–C(=O)–
		H	OH	
134	OTs	OH	H	F_3C–C(=O)–
		H	OH	
135a	NH_2	OH	H	H
135b		H	OH	

A further improvement of the efficacy against resistant strains was reported for the 5″-amino-3′,4′,5″-trideoxy derivatives, the syntheses of which has been published by two different groups (W. K. Woo et al.[178] and H. Saeki et al.[179]). W. K. Woo et al. protected the 3′,4′-hydroxyl groups of *135* via a cyclohexylidine group, acetylated the unprotected hydroxyl functions and removed the cyclohexylidine groups. The liberated 3′,4′-hydroxyl groups were mesylated. Subsequent iodination and reduction yielded 5″-amino-3′,4′,5″-trideoxybutirosin A *118*.

H. Saeki et al. use *136*, as starting material, which is an intermediate in the synthesis of 3′,4′-dideoxybutirosin A. Following the removal of the 3″,5″-cyclohexylidene group, the 5″-position is tosylated and converted to the azide, which gives the 5″-amino function after reduction. The dehydroxylation is accomplished according to the Cohen-Tipson procedure.

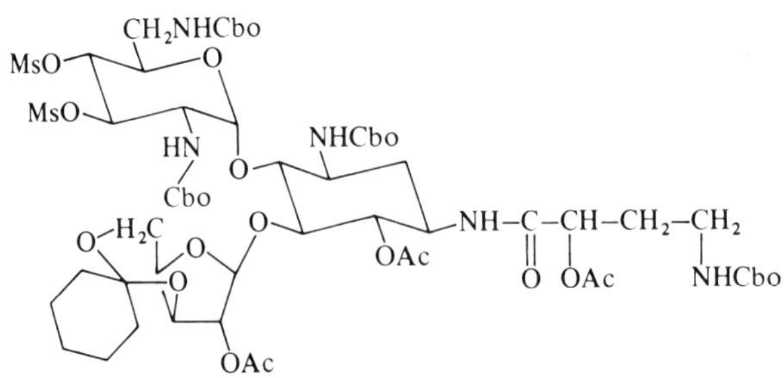

136

Fig. 39.

III.2.3.2 Neomycins, Paromomycins, Lividomycins. Although the tetra- and pentacyclic aminoglycosides have been known for some time (neomycin 1949, paromomycin 1959, lividomycin 1967), only a few chemical modifications have been carried out. Not until after the discovery of the enzymatic inactivation mechanisms were several deoxy compounds and acylated derivatives (HABA, etc.) prepared.

Table 5

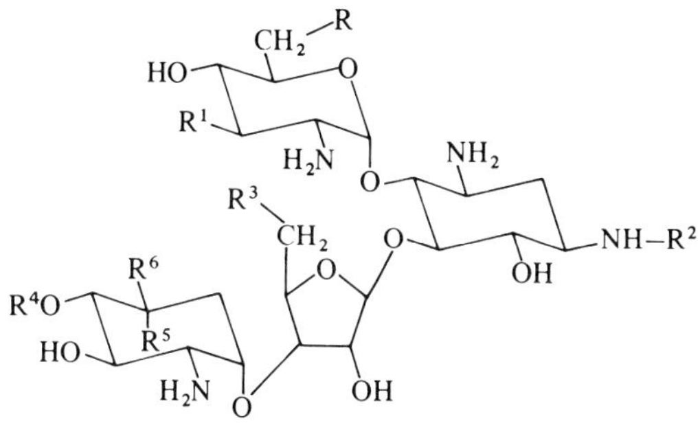

		R	R¹	R²	R³	R⁴	R⁵	R⁶	Ref.
137	6′-N-alkyl-neomycin B	ALKYL-NH	OH	H	OH	H	CH_2NH_2	H	[186]
138	6′-N-aralkylneomycin B	ARALKYL-NH	OH	H	OH	H	CH_2NH_2	H	[186]
139	6‴-N-alkylneomycin B(-paromomycin I)	NH_2 (OH)	OH	H	OH	H	CH_2NH_2	H	[186]
140	6‴-N-aralkylneomycin B (-paromomycin I)	NH_2 (OH)	OH	H	OH	H	CH_2NH_2	H	[186]
141	1-N-HABA-neomycin B	NH_2	OH	HABA	OH	H	CH_2NH_2	H	[198]
142	1-N-acylparomomycin I	OH	OH	ACYL	OH	H	CH_2NH_2	H	[199]
143	6′-Deoxyparomomycin I	H	OH	H	OH	H	CH_2NH_2	H	[191]
144	6′,5″-Dideoxyparomomycin I	H	OH	H	H	H	CH_2NH_2	H	[191]
145	5″-Amino-5″-deoxyparomomycin	OH	OH	H	NH_2	H	CH_2NH_2	H	[194]
146	6′-Amino-6′-deoxylividomycin B (≡3′-Deoxyneomycin B)	NH_2	H	H	OH	H	CH_2NH_2	H	[190]
147	6′-Methylamino-6′-deoxylividomycin B (≡6′-N-methyl-3′-deoxyneomycin B)	CH_3NH	H	H	OH	H	CH_2NH_2	H	[190]
148	6′-(2-Hydroxyethyl)-amino-6′-deoxylividomycin B (≡6′-N-hydroxyethyl-3′-deoxylividomycin B)	$HOCH_2CH_2$–NH	H	H	OH	H	CH_2NH_2	H	[190]

Table 5 (continued)

		R	R^1	R^2	R^3	R^4	R^5	R^6	Ref.
149	5″-Deoxylividomycin B	OH	H	H	H	H	CH_2NH_2	H	192)
150	5″-Deoxylividomycin A	OH	H	H	H	a	CH_2NH_2	H	193)
151	5″-Amino-5″-deoxylividomycin A	OH	H	H	NH_2	a	CH_2NH_2	H	193)
152	6′,5″,6⁗-Triamino-6′,5″,6⁗-trideoxylividomycin A	NH_2	H	H	NH_2	b	CH_2NH_2	H	200)
153	6′,5″-Diamino-6′,5″-dideoxylividomycin B	NH_2	H	H	NH_2	H	CH_2NH_2	H	200)
154	1-N-HABA-lividomycin A	OH	H	HABA	H	a	CH_2NH_2	H	195) 196)
155	6′-Amino-1-N-HABA-6′-deoxylividomycin A	NH_2	H	HABA	OH	a	CH_2NH_2	H	197)

a
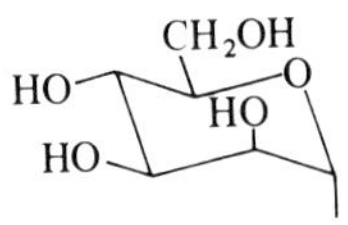

b
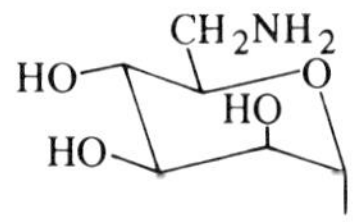

The poly-N-alkyl-[181] and hexa-N-acyl derivatives[182, 183] of neomycin and paromomycin fail to show activity, but the hexa-N-methane-sulfonates[184] and -sulfinates[185] are claimed to be active and less toxic. 6′- and 6‴-N-alkyl- and N-alkylaryl-neomycins and -paromomycins have been prepared from the Schiff's bases by reduction with $NaBH_4$[186]. Some of the alkylaryl compounds show a slightly improved activity; the alkyl derivatives are ineffective. The hexa-N-benzylneomycins[187], prepared from neomycin, aromatic aldehydes and $NaBH_4$, as well as the corresponding paromomycins, synthesized by means of catalytic reduction of the Schiff's bases, have reduced antibacterial activity. The Schiff's bases of paromomycin show in vitro the same activity as the parent compound, obviously because of the ease of hydrolysis.

The controlled synthesis of 2′-N-aralkylparomomycin succeeded through the conversion of paromomycin to the 1,3,2‴,6‴-tetra-N-acetyl derivative, followed by reaction with an aldehyde and reduction of the resulting Schiff's base with $NaBH_4$[189].

3′-Deoxyneomycin *146* (≡6′-amino-6′-deoxylividomycin B) was prepared by the controlled replacement of the 6′-hydroxyl group of lividomycin B by an amino group[190] (Fig. 40). The 4′- and 6′-hydroxyl groups in per-N-benzyloxycarbonyl-lividomycin B was protected by a benzylidine group followed by the acetylation of the remaining hydroxyl groups. Sequential removal of the benzylidine group, controlled tosylation in the 6′-position, introduction of the azido group and reduction gave the 3′-deoxyneomycin. This compound has higher activity against *Pseudomonas sp.* than lividomycin B and it has a broader antibacterial spectrum than neomycin B. Starting with 6′-tosyllividomycin B, the 6′-N-methyl-3′-deoxyneomycin B *147* and 6′-N-(2-hydroxyethyl)-3′-deoxyneomycin B *148* have also been prepared in a similar way.

6′-Deoxy- and 6′-5″-dideoxyparomomycin *143, 144* were prepared from the penta-N-benzyloxycarbonylparomomycin by reaction with tosylchloride, after separation of the two intermediates 6′-tosyl- and 6′,5″-ditosylparomomycin and subsequent reduction[191]. The importance of the 5″-hydroxyl group within the ribose containing aminoglycoside antibiotics (see ribostamycin) has been demonstrated by suitable modifications of lividomycin.

The protection of the 4′- and 6′-hydroxyl groups by treatment with benzaldehyde dimethyl acetal and subsequent tosylation of the 5″-hydroxyl group led, according to a known route (Fig. 27) to 5″-deoxylividomycin B *149*[192]. 5″-Deoxylividomycin A *150* has been prepared similarly via the 4′,6′:2⁗, 3⁗:4⁗,6⁗-triisopropylidene derivative *158*. From the 5″-tosylate *157*, it is also possible to obtain 5″-amino-5″-deoxylividomycin A *151*[193]. Recently the 5″-amino-5″-deoxyparomomycin has been prepared by a synthesis analogous to that of *151*[194].

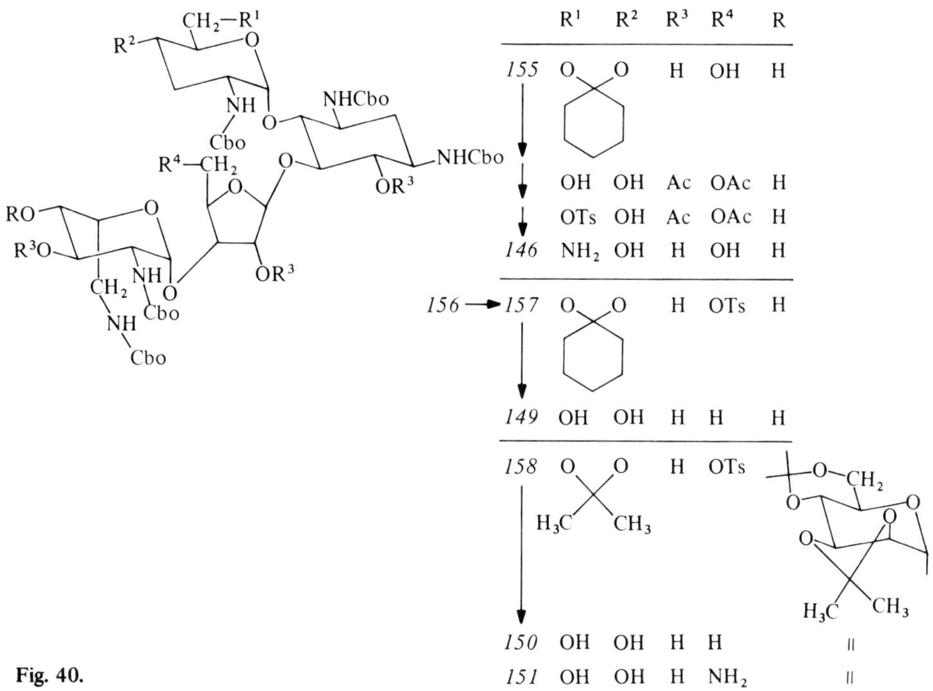

Fig. 40.

The 1-N-HABA derivatives of lividomycin, neomycin and paromomycin have been synthesised according to Umezawa via cyclic carbamates as well as by 1-N-acylation of the 6‴-benzyloxycarbonyl protected compounds. Thereby it was noted that the HABA moiety does not have the same advantageous influence on these aminoglycosides as on the tricyclic aminoglycosides. 1-N-HABA-lividomycin A *154*[195, 196], as well as 6′-amino-1-N-HABA-6′-deoxylividomycin A *155*[197], show no improved activity. 1-N-HABA-neomycin B *141* and C[198] and 1-N-HABA-paromomycin *142*[199], which have about the same activity as the parent coumpounds, have been synthesised by acylation of the corresponding 6′6″-benzyloxycarbonylneomycin or 6‴-benzyloxycarbonylparomomycin through the active ester of (S)-4-amino-2-hydroxybutyric acid.

Lividomycin A has been converted into lividomycin B by oxidative cleavage of the mannosyl moiety with periodate[231]. Lividomycin B can be converted in a similar manner to deneosaminyllividomycin B *119*[180] (Table 4).

III.2.4 4,6-Disubstituted Deoxystreptamine Antibiotics

III.2.4.1 The Kanamycins. Among the aminoglycoside antibiotics compounds of structural type 2c play an important role in antibacterial chemotherapy. Great efforts have, therefore, been made to get new potent compounds of this class by fermentation or by chemical modification.

The 4,6-disubstituted deoxystreptamine aminoglycosides can be divided into kanamycins (Table 6) and gentamicins (Table 8), depending on the structure of the sugar moiety linked to the 6-position of the deoxystreptamine ring.

The persubstituted N-methanesulfonates[184], N methancsulfinates[185a] and Schiff's bases[201] of kanamycin A show the same MIC values as the parent compound due to their ability to hydrolyse under physiological conditions. The $NaBH_4$ reduction of suitably substituted Schiff's bases affords tetra-N-phenylalkyl derivatives of kanamycin A[202], which show activity only in some cases. The first attempted modifications have been the reactions at the 6″-hydroxyl group of kanamycin A. Through protection of the amino groups by preparations of Schiff's bases, acetates or carbamates, the 6″-hydroxyl group is modified to lead to the derivatives, shown in Table 7.

6″-Deoxykanamycin A *159* and 6″-chloro-6″-deoxykanamycin A *160*[203] are prepared via the 6″-tosylate. This tosylate affords 6″-amino-6″-deoxykanamycin A *161* by treatment with sodium azide and subsequent reduction, while the reaction with hydrazine gives 6″-hydrazino-kanamycin A *162*[204]. The corresponding 2-"manno"-compounds are synthesizes from the 2″,6″-ditosyl derivative *191*[205].

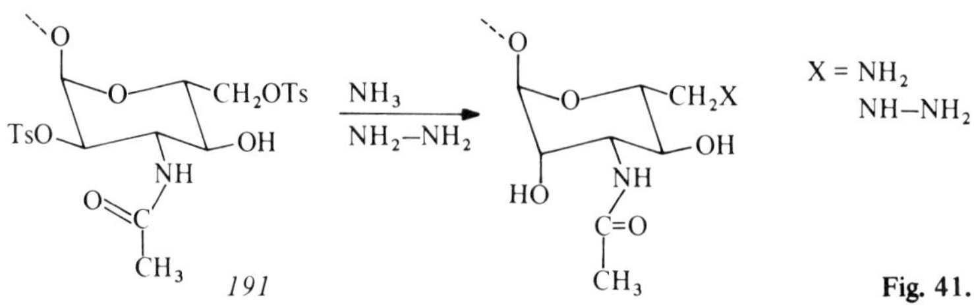

Fig. 41.

The synthesis of kanamycin A-6″-phosphate *163* is achieved by the reaction of N-protected kanamycin A with diphenylchlorophosphate and hydrolytic removal of the phenyl groups[206]. The conversion of the alcoholic function in 6″-position to a

Table 6

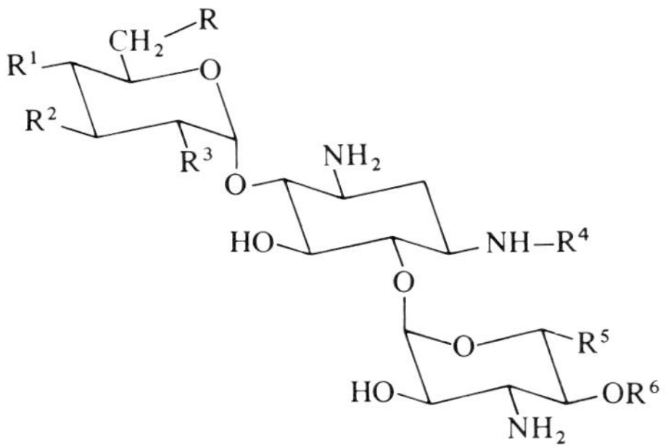

		R	R^1	R^2	R^3	R^4	R^5	R^6	Ref.
	6″-Modified Kanamycin A	NH_2	OH	OH	OH	H	see Table 7	H	203–207)
168	4″,6″-O-benzylidine-kanamycin A	NH_2	OH	OH	OH	H	$-H_2CO-CH(Ar)-O-$		208)
169	3′-Deoxykanamycin A	NH_2	OH	H	OH	H	CH_2OH	H	210)
170	3′-O-methylkanamycin A	NH_2	OH	OCH_3	OH	H	CH_2OH	H	211)
171	3′-Amino-3′-deoxykanamycin A	NH_2	OH	NH_2	OH	H	CH_2OH	H	209)
172	3′-Amino-3′-deoxy-2′-"manno"-kanamycin A	NH_2	OH	NH_2	OH	H	CH_2OH	H	209)
173(173a)	4′-Deoxykanamycin A (B)	NH_2	H	OH	OH (NH_2)	H	CH_2OH	H	214, 215, 215c)
174	3′-Deoxykanamycin B	NH_2	OH	H	NH_2	H	CH_2OH	H	212)
175	3′,6″-Dideoxykanamycin B (6″-Deoxytobramycin)	NH_2	OH	H	NH_2	H	CH_3	H	213)

Table 6 (continued)

		R	R^1	R^2	R^3	R^4	R^5	R^6	Ref.
176	3′,4′-Dideoxykanamycin B (DKB)	NH_2	H	H	NH_2	H	CH_2OH	H	216)
177	6′-N-methylkanamycin B	$NHCH_3$	OH	OH	NH_2	H	CH_2OH	H	217)
178	3′,4′-Dideoxy-6′-N-methyl-kanamycin B	$NHCH_3$	H	H	NH_2	H	CH_2OH	H	217)
179	1-N-HABA-kanamycin A (BB-K8, amikacin)	NH_2	OH	OH	OH	HABA	CH_2OH	H	219, 219a)
180	1-N-acylkanamycin A	NH_2	OH	OH	OH	ACYL	CH_2OH	H	225, 226, 232)
181	1-N-acylkanamycin B	NH_2	OH	OH	NH_2	ACYL	CH_2OH	H	226)
182	6′-N-HABA-kanamycin A	HABA	OH	OH	OH	H	CH_2OH	H	218)
183	1-N-HABA-3′,4′-dideoxy-kanamycin B	NH_2	H	H	NH_2	HABA	CH_2OH	H	166)
184	1-N-HABA-3′-deoxykanamycin B (1-N-HABA-tobramycin)	NH_2	OH	H	NH_2	HABA	CH_2OH	H	222)
184a	I-N-HABA-3′-deoxykanamy-cin A	NH_2	OH	H	OH	HABA	CH_2OH	H	216a)
185	2′-N-HABA-3′,4′dideoxy-kanamycin B	NH_2	H	H	NH–HABA	H	CH_2OH	H	224)
186	1,2′-N-bis-HABA-3′,4′-dideoxykanamycin B	NH_2	H	H	NH–HABA	HABA	CH_2OH	H	224, 224a)
187	1-N-HABA-6′-N-alkylkana-mycin A (6′-N-alkylamikacin)	ALKYL-NH	OH	OH	OH	HABA	CH_2OH	H	227)
188	1-N-HABA-3′,4′-dideoxy-6′-N-methylkanamycin B	H_3C-NH	H	H	NH_2	HABA	CH_2OH	H	223)
189	1-N-acyl-3′,4′-dideoxy-6′-N-methylkanamycin B	H_3C-NH	H	H	NH_2	ACYL	CH_2OH	H	223)
190	1-N-(ω-amino-2-hydroxy-alkyl)-kanamycin A (B)	NH_2	OH	OH	OH (NH_2)	$H_2N-CH_2-CH_2-CH_2-CH(OH)-$	CH_2OH	H	230)

Table 7

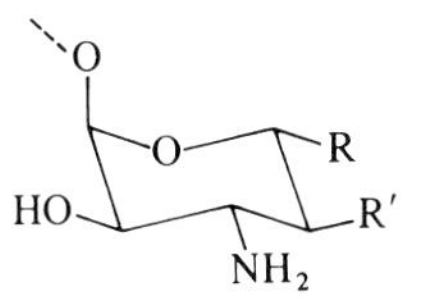

	R	R'	
159	$-CH_3$	$-OH$	
160	$-CH_2-Cl$	$-OH$	
161	$-CH_2-NH_2$	$-OH$	Also as "manno"-link
162	$-CH_2-NH-NH_2$	$-OH$	Also as "manno"-link
163	$-CH_2-O-P(OH)_2$ (→O)	$-OH$	
164	$-COOCH_3$	$-OH$	
165	$-COOH$	$-OH$	
166	$-CONH_2$	$-OH$	
167	$-CH_2O$	$-O$	

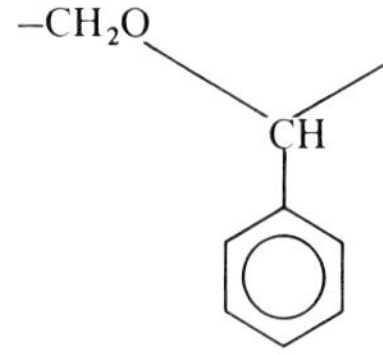

carbonylic function in the kanamycin molecule is accomplished through the following steps: tritylation of the 6''-hydroxyl group, acetylation of the remaining hydroxyl functions, removal of the trityl group, oxidation with potassium permanganate and deprotection[207]. The antibacterial activity is reduced by a factor of 5–10 in the sequence $-CO_2H > -CO_2CH_3 > CONH_2$.

4''-6''-0-benzylidene-kanamycin A *168* has no antibacterial activity[208].

The importance of the 3'- and 4'-hydroxyl groups for the inactivation in resistant organisms gave impetus to many variations of these positions. 3'-Amino-3'-deoxykanamycin A *171* and 3'-amino-3'-deoxy-2'-mannokanamycin A *172* are prepared from tetra-N-acetylkanamycin A as follows[209]: Oxidative cleavage of the 6'-aminoglucose moiety to yield the dialdehyde *192,* and subsequent cyclisation with nitromethane to compound *193.* Reduction with Raney Nickel catalyst leads to the isomers *194* and *195.*

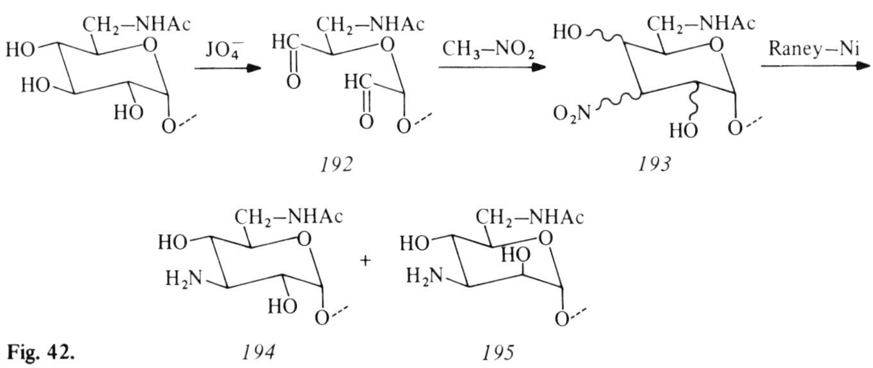

Fig. 42.

3′-Deoxy- and 3′-0-methylkanamycin A *169, 170*[210, 211] are synthesised via glycosylation of the substituted deoxystreptamine derivatives through the appropriate 3′-deoxy- or 3′-0-methylsugars. For the synthesis of 3′-deoxykanamycin B *174* (identical with the microbiologically produced tobramycin), the starting material is penta-N-ethoxycarbonylkanamycin B *196,* which is converted into the 3′,4′: 4″,6″-di-0-cyclohexylidene derivative *197* according to Bissett's procedure[142]. Controlled cleavage of the 3′,4′-cyclohexylidene group and reaction with p-toluenesulfonylchloride in pyridine affords the 3′-tosyl compound *198* and the 3′,4′-ditosyl derivative. The tosylate *198* is converted into 3′-deoxykanamycin B[212], which displays a substantial improvement in activity when compared to kanamycin B.

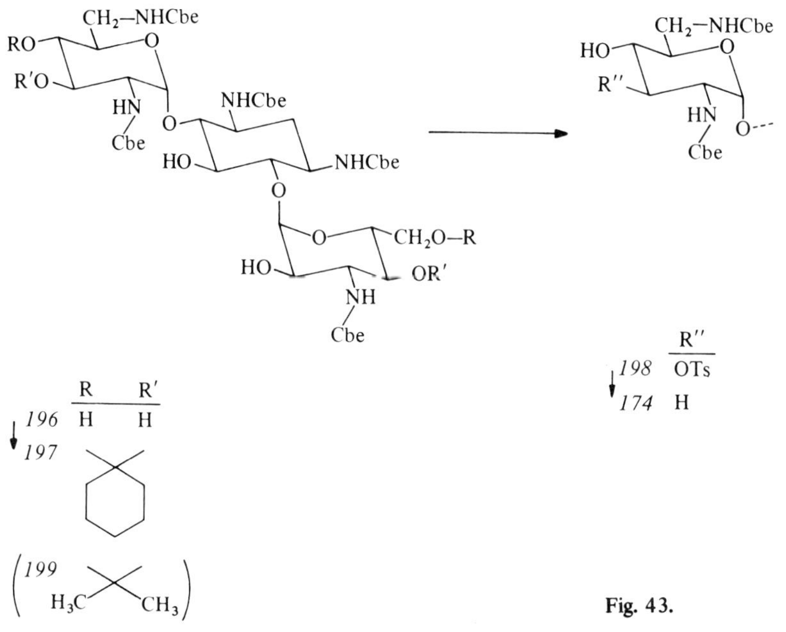

Fig. 43.

The synthesis of 3′,6″-dideoxykanamycin B *175,* identical with 6″-deoxytobramycin, is achieved by conversion of penta-N-acetylkanamycin B into the 3′,6″-diiodokanamycin B and reduction[213].

4′-deoxykanamycin A *173* is readily obtained using glycoside synthesis or by chemical modification of kanamycin A[214]. The dehydroxylation, starting with 6′-N-benzyloxycarbonylkanamycin A, is accomplished in two ways[215]: protection of the amino groups with an ethyloxycarbonyl residue, hydrogenolytic removal of the 6′-benzyloxycarbonyl group, protection of the 4′- and 6′-position by treatment with phenylchloroformate, protection of the remaining hydroxyl functions to yield *200,* cleavage of the 4′,6′-cyclic carbamate to give *201,* and sequential mesylation (*202*), iodination (*203*), elimination and hydrogenation to yield the 4′-deoxykanamycin A.

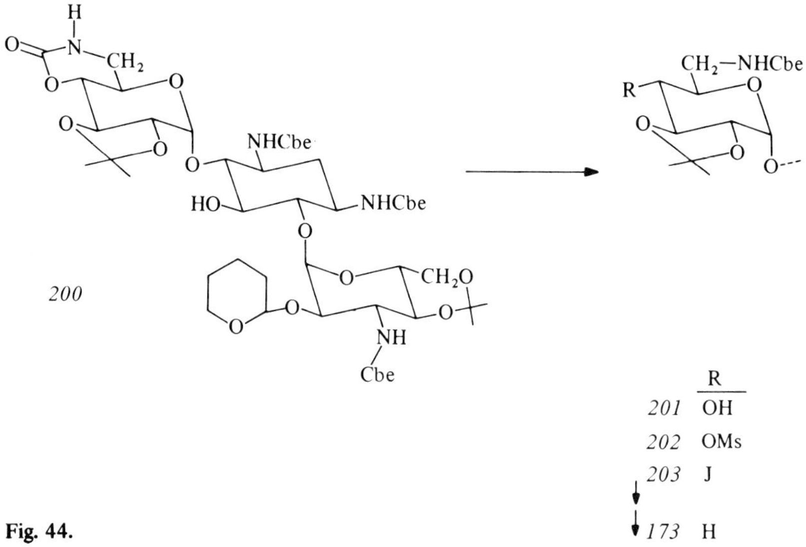

200

	R
201	OH
202	OMs
203	J
173	H

Fig. 44.

The alternative way started with an acetylation of 6′-N-benzyloxycarbonylkanamycin A, during which the hindered 5-hydroxy group remained unchanged. After deprotection of the 6′-amino group by hydrogenolysis, an O → N acetyl migration from 4′-OH to 6′-NH_2 occurred. The liberated 4′-hydroxyl group is removed as previously described. Recently the 4′-deoxykanamycin B *173* was prepared through a modified synthesis[215c)].

3′,4′-Dideoxykanamycin B (Dibekacin, DKB) *176* is synthesised using the Tipson-Cohen procedure[216)]. The intermediate *199* (Fig. 43) is benzoylated in the 2″-position and the 3′- and 4′-hydroxyl groups are deprotected. Subsequent 3′,4′-dimesylation, iodination and reduction give DKB in about 5% yield. A modified procedure for the synthesis of DKB has been published recently[216d)]. DKB is effective against such kanamycin-resistant organisms which phosphorylate the 3′-hydroxyl group of the kanamycins. To prevent a second inactivation mechanism, viz. the 6′-N-acetylation, Umezawa has converted kanamycin B and DKB into the 6′-benzyloxycarbonylderivatives, which by reduction with $LiAlH_4$ afford 6′-N-methylkanamycin B *177* and 3′,4′-dideoxy-6′-N-methylkanamycin B *178*[217)]. The 6′-N-methyl compounds are more potent against resistant strains.

For the introduction of the HABA moiety in kanamycin advantage can be taken of the small differences in the reactivity of the amino groups[218)]. Acylation leads in varied amount to isomeric compounds, which are separated by chromatography on acidic exchange resins.

1-N-HABA-kanamycin A *179* (Amikacin, BB-K8) is prepared from kanamycin A in 10% yield[219, 219a)] by treatment of 6′-benzyloxycarbonylkanamycin A with the active ester of 2-hydroxy-4-(benzyloxycarbonylamino) butyric acid *204* or with *204a*[220)] followed by deprotection.

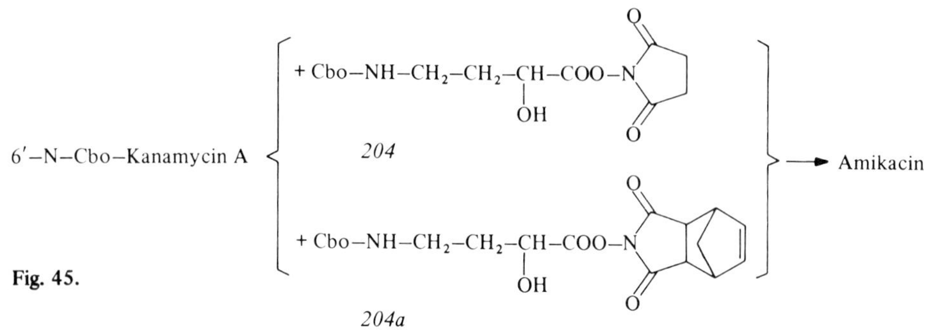

Fig. 45.

The reaction of *204* with kanamycin A and subsequent deprotection affords 6′-N-L-(-)-HABA-kanamycin A (BB-K6) *182*. If the 6′- and 1-positions are previously protected the 3-N-L-(-)-HABA-kanamycin A (BB-K29) is isolated. Following the additional protection of the 3-position, the reaction with the active ester leads to 3″-N-L-(-)-HABA-kanamycin A (BB-K 11)[218]. These isomers are thirty to a hundred times less active than BB-K8 (Amikacin).

Amikacin is more potent than kanamycin A and superior to BB-K19 (D,L-HABA) and BB-K31 (D(+)-HABA). Amikacin is the only important semisynthetic aminoglycoside on the market. In this connection two possibilities to synthesise the HABA moiety should be described. One of these syntheses starts with butyrolactone and uses the steps shown in Fig. 46a. Another method utilizes the 1,3-dipolar addition of the nitrone ester *205* to ethyl acrylate (Fig. 46b).

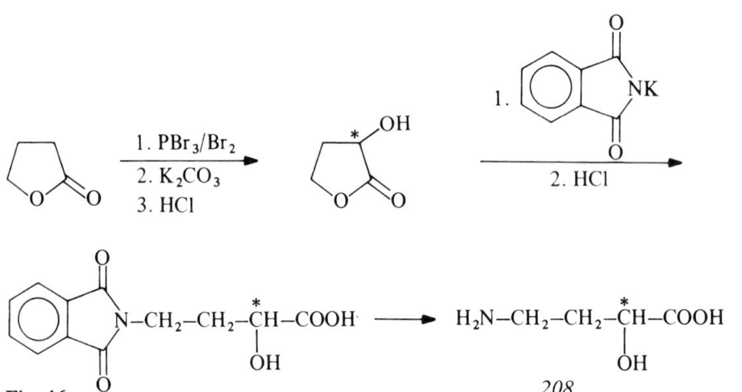

Fig. 46a.

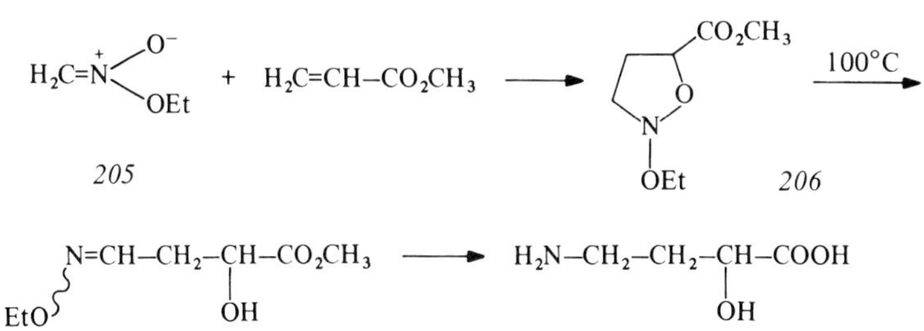

Fig. 46b. *207* *208*

205 *206*

For the synthesis of kanamycin B derivatives such as 1-N-HABA-kanamycin B *181,* 1-N-HABA-3′,4′-dideoxykanamycin B *183*[219 a, 166] and 1-N-HABA-3′-deoxykanamycin B *184* (identical with 1-N-HABA-tobramycin[222]), one has to bear in mind that kanamycin B has one more amino group than kanamycin A. Treatment with benzyloxycarbonylchloride or tert-butyloxycarbonylazide protects first the 6′-amino group and subsequently the 2′-amino function. Reaction of the 2′,6′-N-protected kanamycin B derivatives with the protected active ester of HABA leads to the 1-acylated compounds[223].

The compounds *181, 183* and *184* are very effective against such kanamycin resistant strains, which produce phosphotransferases I and II and nucleotidyl-transferase. The following isomeric HABA-3′,4′-dideoxykanamycin B derivatives have been synthesised[224]: 3-,2′- and 3″-HABA-DKB and 1,2′- or 3,2′-bis-HABA-DKB. 1,2′-bis-HABA-DKB is a little less active than 1-N-HABA-DKB, whereas the other derivatives have very low activity.

Since it was known that the HABA moiety causes a substantial improvement in activity, a great number of HABA analogues have been prepared and introduced in different aminoglycosides.

D,L-, L- and D-isoserylkanamycin A[225] show about 80% of the activity of BB-K8. A comparison of propionyl- or valerylkanamycin A (BB-K101 or BB-K23) with propionyl- or valerylkanamycin B (BB-K122 or BB-K33) shows that the elongation of the side chain to more than 4 C-atoms leads to a decrease of activity[226, 232].

1-N-HABA-6′-N-alkyl-kanamycin A *187* (=6′-N-alkylamikacin)[227] and 1-N-HABA-3′,4′-dideoxy-6′-N-methylkanamycin B *118*[228] contain additional modifications, which are supposed to prevent enzymatic inactivation.

1-N-HABA-6′-N-(methyl or ethyl)-kanamycin A is prepared from 6′-N-tert. butyloxycarbonyl-BB-K8, an intermediate of the BB-K8 synthesis. Subsequent benzyloxycarbonylation in 3′- and 3″-position and removal of the 6′-N-protecting group is followed by a reductive alkylation by formaldehyde or acetaldehyde and sodium borohydride. The controlled acylation of 6′-N-methylkanamycin A and B in 1-N-position using the active ester of HABA is also described[229].

1-N-acyl-3′,4′-dideoxy-6′-N-methylkanamycin B *189* is obtained through the1-N-acylation[223] of 3′,4′-dideoxy-6′-N-methylkanamycin B *178*[217]. The acyl groups in these cases are HABA and the corresponding propionyl- and valeryl moieties. All the derivatives are very potent and also active against organisms which are inactivate in the 3′,6′- and 2″-positions of the aminoglycoside antibiotics.

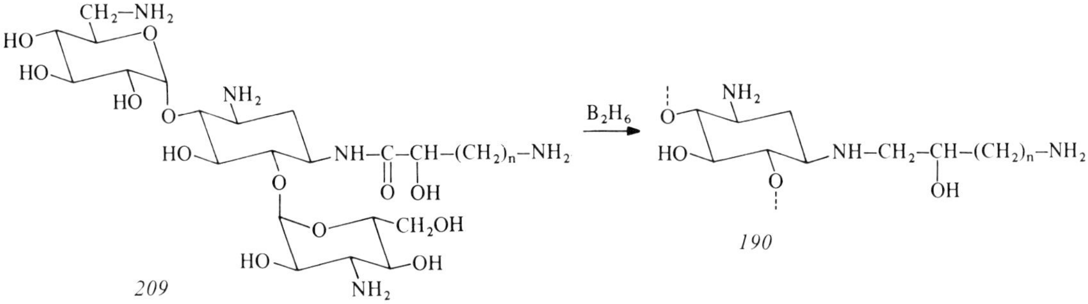

209
(n=2: Amikacin)

190

Fig. 47.

The reduction of the described 1-N-[(S)-ω-amino-2-hydroxyacyl]-kanamycin derivatives by treatment with diborane leads to the corresponding 1-N-[(S)-ω-amino-2-hydroxyalkyl]-kanamycins *190*[230]. 1-N-[(S)-4-amino-2-hydroxybutyl]-kanamycin A has been shown to possess activity similar to that of amikacin[230a].

Instead of the utilization of compound *209*, kanamycin can also be acylated in 1-position with $HO_2C{-}CH(OH){-}(CH_2)_{n-1}{-}CONH_2$ to give *210*. Compound *210* was then reduced to the desired amino derivative.

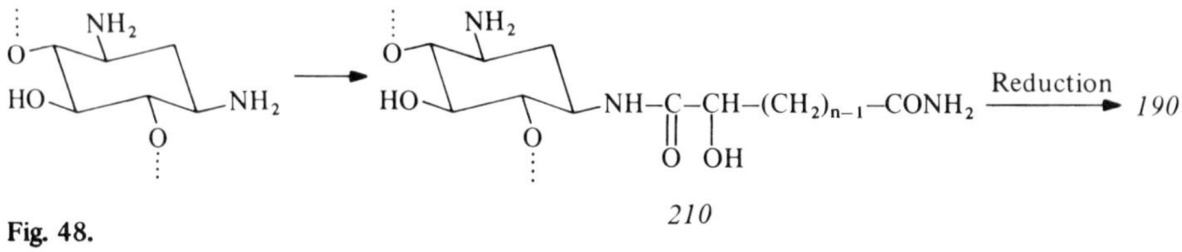

Fig. 48.

III.2.4.2 Gentamicins and Related Compounds. Gentamicin C, a mixture of the three components C_1, C_2, C_{1a}, is by far the most important aminoglycoside in modern therapy. A great number of modified gentamicins have been prepared using chemical or biological (Chap. III.3) methods. The results of these efforts afford an insight into the structure-activity relations of the aminoglycosides.

Gentamicin C_2 and C_{1a} react with 5 moles of an aromatic aldehyde to give tetraarylidene-oxazolidine compounds (*242*), which are similar in activity to the parent compounds, have lower serum levels but longer half-lives. The reaction of these Schiff's bases with sodium borohydride furnished the inactive tetrabenzylgentamicin C_2 and C_{1a} *243*. The oxazolidine ring is opened by the reduction[233, 234].

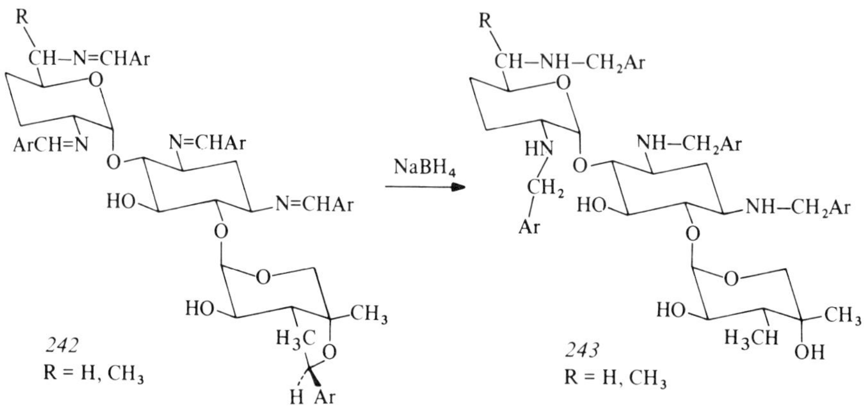

Fig. 49.

6′-N-arylmethylgentamicin C_{1a} and $-C_2$ are prepared in a similar manner, starting with a N-protected gentamicin bearing a free 6′-amino group[235].

T. R. Beattie[236] has described several modifications in 2″- and 5-position (*211–222*) of the gentamicins C_1, C_2 and C_{1a}, all of which are synthesised by treatment

Table 8

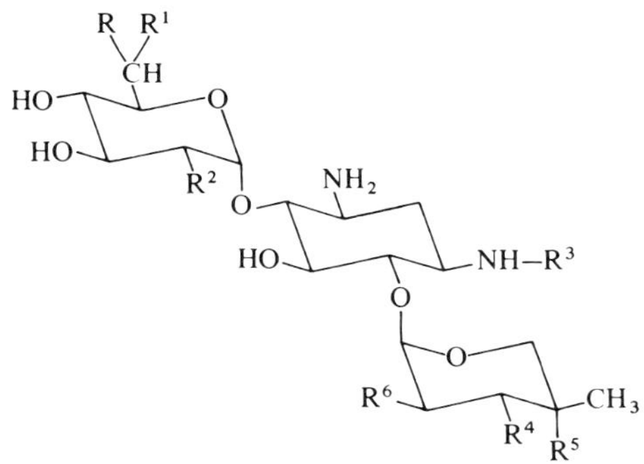

No.	Compound	R	R^1	R^2	R^3	R^4	R^5	R^6	Ref.
211	2″-OH } modified gentamicin C_1,C_2,C_{1a}		see Table 9						236)
222	5-OH } modified gentamicin C_1,C_2,C_{1a}		see Table 9						236)
223	6′-N-arylmethylgentamicin $C_{1a}(C_2)$	H,(CH_3)	NH–CH_2–Ar	NH_2	H	$NHCH_3$	OH	OH	235)
224	4″-Deoxygentamicin C_1	CH_3	$NHCH_3$	NH_2	H	$NHCH_3$	H	OH	237)
225	2″-Epi-gentamicin C_1	CH_3	$NHCH_3$	NH_2	H	$NHCH_3$	OH	OH	238)
226	2″-Deoxygentamicin C_2	CH_3	NH_2	NH_2	H	$NHCH_3$	OH	H	238, 239)
227	2″-Deoxy-3″-demethylamino-2″-methylaminogentamicin C_2	CH_3	NH_2	NH_2	H	H	OH	$NHCH_3$	238, 239)
228 *228a*	3″-N-demethylgentamicin $C_1(C_2)$	CH_3	$NHCH_3$ (NH_2)	NH_2	H	NH_2	OH	OH	250)
229	6′,3″-N-demethylgentamicin C_1	CH_3	NH_2	NH_2	H	NH_2	OH	OH	250)
230	3″-N-demethylsagamicin	H	$NHCH_3$	NH_2	H	NH_2	OH	OH	251)
231	1-N-HABA-gentamicin C_1,(C_2),C_{1a}	CH_3 (CH_3) H	$NHCH_3$ (NH_2) NH_2	NH_2	HABA	$NHCH_3$	OH	OH	241, 242, 248)
232	1-N-acylgentamicin C_2	CH_3	NH_2	NH_2	ACYL	$NHCH_3$	OH	OH	242)
233	2′-N-HABA-gentamicin C_1	CH_3	$NHCH_3$	NH-HABA	H	$NHCH_3$	OH	OH	241)
234	1-N-HABA-sagamicin	H	$NHCH_3$	NH_2	HABA	NH_2	OH	OH	244)
235	1-N-acylsagamicin	H	$NHCH_3$	NH_2	ACYL	NH_2	OH	OH	245)
236	5′-Epi-gentamicin C_{1a}	H	NH_2	NH_2	H	$NHCH_3$	OH	OH	257)
237	5-Epi-gentamicin etc.	}	see Fig. 54						253)
238	5-Epi-amino-5-deoxygentamicin etc.	}	see Fig. 54						253)

Table 8a.

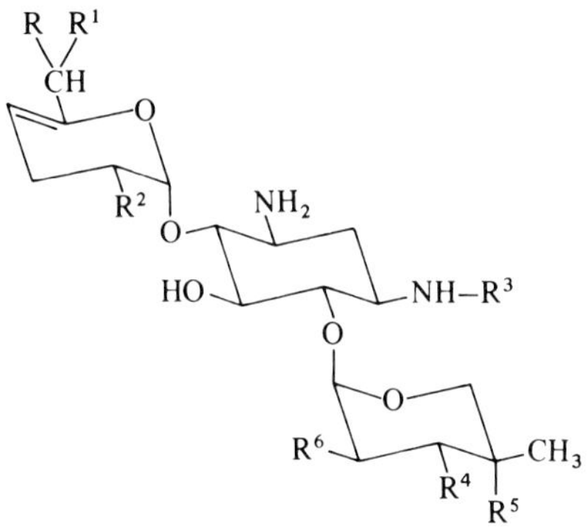

		R	R^1	R^2	R^3	R^4	R^5	R^6	Ref.
239	1-N-acylsisomicin	H	NH_2	NH_2	ACYL	$NHCH_3$	OH	OH	249)
240	1-N-acylverdamicin	CH_3	NH_2	NH_2	ACYL	$NHCH_3$	OH	OH	249)
241	1-Ethylsisomicin (Sch 20569, netilmicin)	H	NH_2	NH_2	C_2H_5	$NHCH_3$	OH	OH	256)

of benzyloxycarbonyl-protected gentamicins with the appropriate acylating reagents (Table 9). There are no available data on the antibacterial activity of these compounds.

Table 9

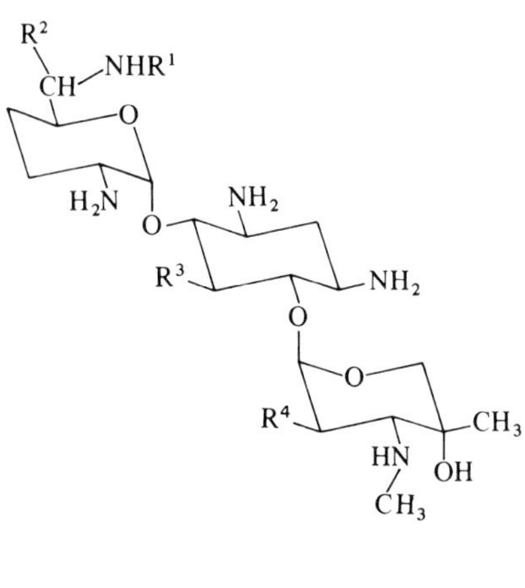

Gentamicin	C_1	C_2	C_{1a}
R^1	CH_3	H	H
R^2	CH_3	CH_3	H

	R^3	R^4
211	$-OH$	$-OCH_3$
212	$-OH$	$=O$
213	$-OH$	$-OSO_2CH_3$ (Epi ~)
214	$-OH$	$-NH_2$ (Epi ~)
215	$-OCONHC_2H_5$	$-OH$
216	$-OCONH_2$	$-OH$
217	$-OSO_2CH_3$	$-OH$
218	$-OH$	$-OCONHCH_3$
219	$-OH$	$-OCONH_2$
220	$-OCH_3$	$-H$
221	$-OCONH_2$	$-OCONH_2$
222	$-OH$	$-$Epi$-$OH

4″-Deoxygentamicin C_1 *224* is obtained by reduction of penta-N-benzyloxycarbonyl-tri-O-acetylgentamicin C_1 with sodium in liquid ammonia[237]. Compound *224* is inactive. Gentamicin C_1 and gentamine C_1 are isolated as by-products.

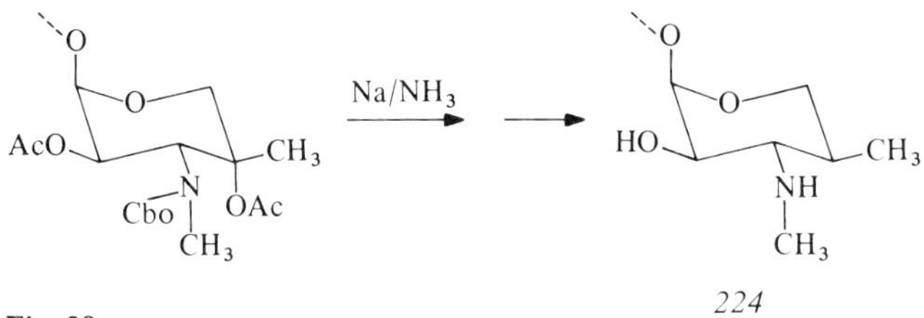

Fig. 50.

P. J. L. Daniels et al.[238] synthesised 2″-epi-gentamicin C_1 *225* and 2″-deoxygentamicin C_2 *226*. Penta-N-acetylgentamicin C_1 is mesylated at the 2″-hydroxyl group. Through solvolysis of the mesylate by participation of the vicinal trans acetamido group, epimerisation at the 2″-C atom takes place.

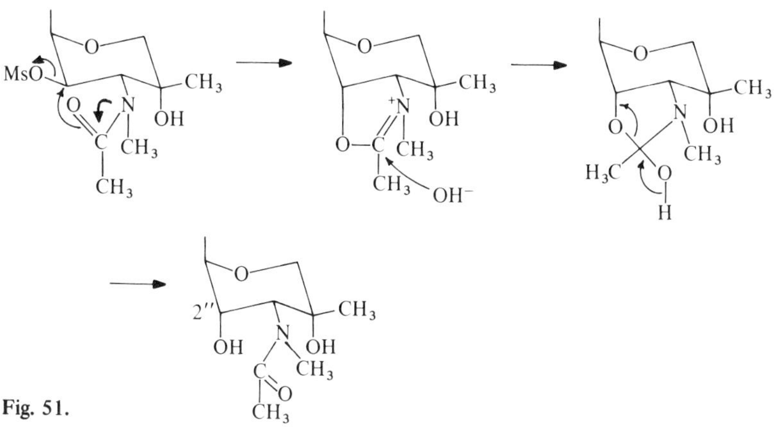

Fig. 51.

The synthesis of 3′-epi-seldomycin 5 from seldomycin 5 (*36*) is achieved by a reaction sequence similar to that shown in Fig. 51. Reaction of the N-protected seldomycin 5 with O-nitrobenzene-sulfonylchloride gives in one step the epicyclic carbamate *244*, which is then hydrolyzed to the 3′-epimer[254] (Fig. 52). The synthesis of 3′-deoxyseldomycin 5 is accomplished by known deoxygenation procedures[255]. Both compounds show, as expected, an improved antibacterial activity against resistant strains which phosphorylate the 3′-hydroxyl group.

The synthesis of 2″-deoxygentamicin C_2 *266* is shown in Fig. 53. During the isolation of *245* a rearrangement takes place to the epi-imino compound *246*. Reduction leads to the inactive main product 2″-deoxy-3″-demethylamino-2″-methylaminogentamicin C_2 *227*, together with 2″-deoxygentamicin C_2 *226*. An alternative method for the synthesis of *226* from *245* is described in a patent application[239].

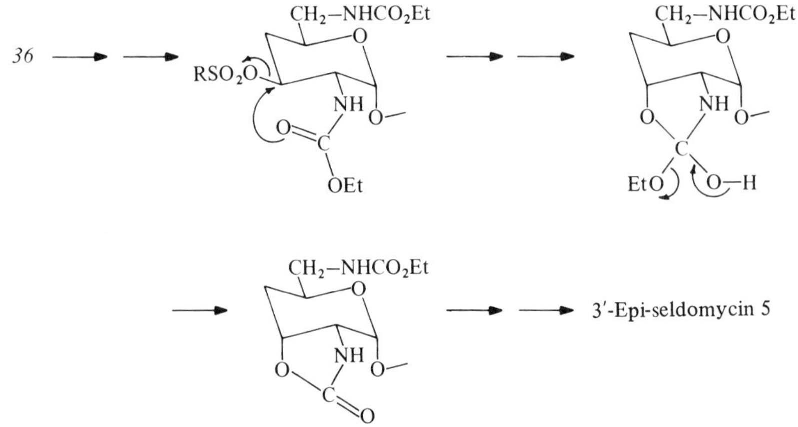

Fig. 52. 244

The alkylthio compound *247* is used as an intermediate, which is converted into *226* through desulfurisation with Raney-Nickel. While the removal of the 2″-hydroxyl group did not affect the activity, epimerisation at this position caused a loss of activity.

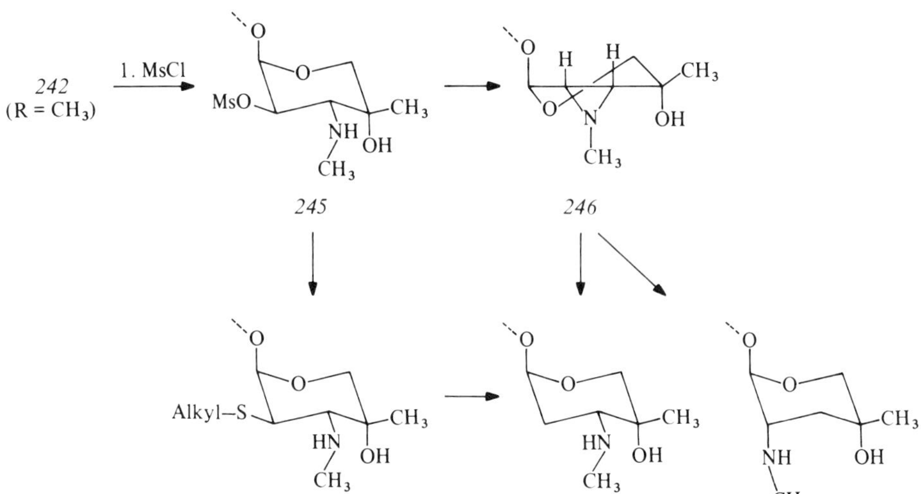

Fig. 53. 247 226 227

The synthesis of 6′-amino-6′-deoxygentamicin A *248* from gentamicin A *43* demonstrates of the influence of the 6′-amino group[240)] to provide an increase in activity. *248* is synthesised by O-acetylation of 6′-trityl-penta-N-benzyloxycarbonyl-gentamicin A, detritylation and conversion of the 6′-hydroxyl group via the tosylate and the azido compound into the 6′-hydroxyl group via the tosylate and the azido compound into the 6′-amino derivative.

A mixture of 1-N- and 2′-N-HABA-gentamicin C_1[241] is obtained by treatment of gentamicin C_1 with the active ester of HABA. In a similar way 1-N-HABA-gentamicin C_2 *231*, 1-N-[(S)-3-amino-2-hydroxypropionyl]-gentamicin C_2 *232* and 1-N-[(S)-5-amino-2-hydroxyvaleryl]-gentamicin C_2 *232*[242] are also synthesised. For the synthesis of 1-N-HABA-gentamicin C_{1a} the 6′-amino group is first protected[243]. The synthesis of 1-N-HABA-sagamicin (XK-61-2) *234*[244] or 1-N-[(S)-3-amino-2-hydroxypropionyl]-sagamicin *235*[245] succeeds after protection of the 2′-amino function. Without blocking the 2′-amino group, a mixture of different and manifold substituted HABA-sagamicin derivatives is obtained[246]. The synthesis of 1-N-HABA derivatives of the other gentamicin analogues *37–41, 43–58* (Fig. 8, 9) through acylation of the protected parent compound is covered in a patent application[247]. An improved synthesis of 1-N-HABA-gentamicin C_1 and 1-N-[(S)-amino-2-hydroxypropionyl]-gentamicin C_1 is described by P. J. L. Daniels et al.[248]. Daniels utilises the different reactivities of the amino groups of gentamicin C_1, which react with trifluoroethylacetate in the following order: $2' > 3 > 1 > 6' > 3''$. According to this order, the 1-amino groups is acylated after protection of the 2′- and 3-amino function as trifluoroacetamides.

Since the selectivity of the N-acylation is pH-dependent[249], optimal reaction conditions can be found through controlled protonation of the amino groups, which can be achieved by controlled addition of triethylamine to the solution of a neutral gentamicin salt. The active ester of HABA, for instance, reacts selectively with the 1-amino group after the addition of one equivalent of triethylamine to a gentamicin salt.

The 1-N[(S)-ω-amino-2-hydroxyacyl]-gentamicins so far known display lower activity compared to gentamicin, but are good for resistant strains of bacteria.

3″-N-demethylgentamicin C_1 *228*, -C_2 *228a* and 6′,3″-N-demethyl-gentamicin C_1 *229* are prepared by oxidative demethylation, e.g. with iodine as oxidant[250]. Reaction of sagamicin (antibiotic XK-62-2) *42* under similar conditions leads to 3″-N-demethylsagamicin *230* and gentamicin C_{1a} *39*[251]. Gentamicin C_{1a} can be converted into sagamicin by the treatment with carbon disulfide to yield 6′-N-dithiocarbamylgentamicin C_{1a}, followed by reduction[252]. Variations in the deoxystreptamine moiety of numerous representatives of the gentamicin-type compounds have been described in a patent application[253]. The suitably protected starting compound was converted into the 5-epicompound *237*. Subsequent reaction with sodium azide and reduction led to the 5-epi- amino-5-deoxyderivative *238*.

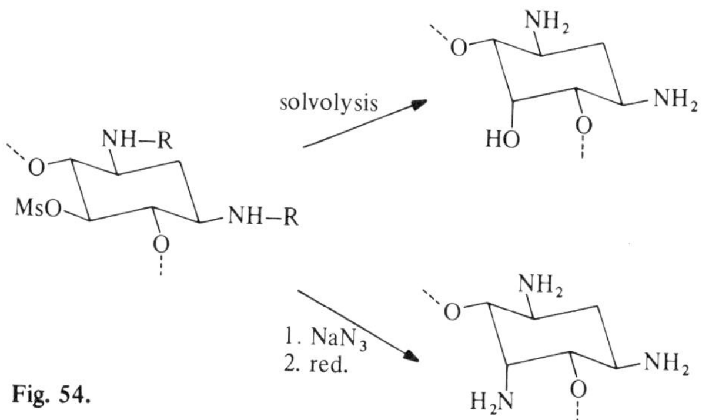

237

238

Fig. 54.

Until now sisomicin and verdamicin have been relatively little modified. The 1-N-acyl derivatives could be obtained under controlled pH conditions. They show activity comparable to that of the parent compound[249]. 1-ethyl-sisomicin (netilmicin) *241* was obtained through reaction with an excess of acetaldehyde and sodium borohydride[256]. Netilmicin shows an improved antibacterial spectrum. The hydrogenation of sisomicin gave 5'-epi-gentamicin C_{1a} *236* (dihydrosisomicin)[257]. This simple conversion caused a substantial loss of activity.

The chemical modifications of the compounds shown in the Figs. 11–18 will not be reported. These compounds and their derivatives will play, if at all, a minor role in the therapy.

III.3 Biosynthesis and Biochemical Modifications of the Aminoglycosides

A comprehensive review on the biosynthesis of aminocyclitol antibiotics was published by K. L. Rinehart and R. M. Stroshane[258]. Most of the biosynthetic studies have been directed to the largest and most important group, the deoxystreptamine-containing antibiotics. The biosynthesis was followed by ^{14}C- or ^{13}C-isotope labelling of subunits and determination of the level of activity in isolated intermediates and in final products by degradation, NMR-spectroscopy and mass spectrometry.

Glucose and glucosamine are precursors of aminocyclitol antibiotics. The specific labelling in some of the intermediate compounds suggests a complicated mechanism of biosynthesis, involving intermolecular rearrangements of the molecule in some cases. Nitrate is the best nitrogen source for kanamycin production. Amino acids, with the exception of glycine, are poor nitrogen sources[259].

The sequence in which the subunits are joined to pseudodisaccharide, pseudodisaccharide, pseudotrisaccaride or higher pseudooligosaccharide is still unclear. Some information is available on the final biosynthetic steps for neomycin and the gentamicin-complex. Phosphoamido neomycins have been isolated and their enzymatic conversion into neomycins has been studied. The location of the phosphoamido group is unknown[260, 261].

The components of the gentamicin complex contain C-methyl and N-methyl substituents, the source of which seems to be mainly due to methylation by L(–) methionine. The level of labelling does not completely support this argument. Reversible enzymatic reactions such as transmethylation, transamidation and deoxygenation are assumed[262].

The biosynthesis of paromomycin, ribostamycin, butirosin, kanamycin and sisomicin proceeds in a manner similar to that of neomycin from deoxystreptamine, but it differs in some details[258].

Some mutants have been used during biosynthesis to incorporate modified deoxystreptamine derivatives. In such so called "mutasynthesis"[258] or "mutational biosynthesis"[263], new aminoglycoside antibiotics can be prepared. A series of analogues of 2-deoxystreptamine has been used in different 2-deoxystreptamine-negative mutants to determine whether they are converted to antibiotics. Only limited substitution at the 2-position (streptamine, epi-streptamine, 1-N-methyldeoxystreptamine) leads to bioconversion to active antibiotics, indicating a rigid requirement for the 1,3-diamino function and the hydroxyl group configuration as in 2-deoxystreptamine[264, 265].

A 2-deoxystreptamine-requiring mutant from *Streptomyces fradiae* converts streptamine and epistreptamine into the hybrimycins *17a–f* (Fig. 4)[23]. A mutant from *Micromonospora inyoensis* produces new sisomicin analogues, the mutamycins, from streptamine and 2,5-dideoxystreptamine[266].

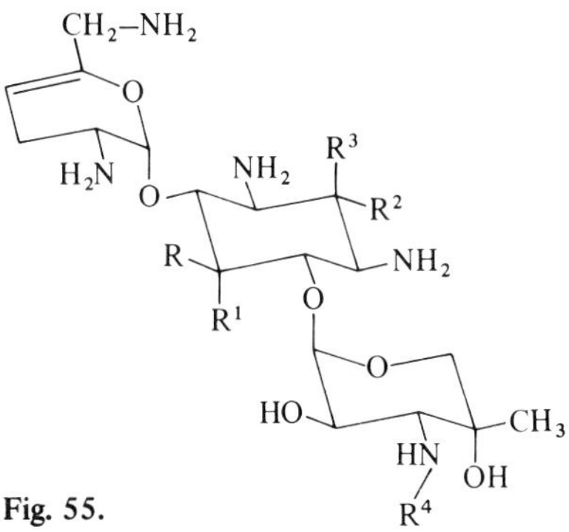

Fig. 55.

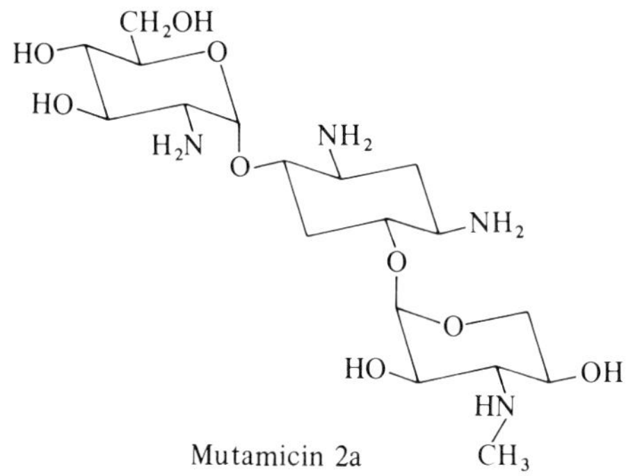

Mutamicin 2a

	R	R^1	R^2	R^3	R^4
Sisomicin	OH	H	H	H	CH_3
Mutamicin 1	OH	H	OH	H	CH_3
Mutamicin 1a	OH	H	OH	H	$COCH_3$
Mutamicin 1b	OH	H	OH	H	H
Mutamicin 2	H	H	H	H	CH_3
Mutamicin 4	OH	H	H	OH	CH_3
Mutamicin 5	NH_2	H	H	H	CH_3
Mutamicin 6	H	OH	H	H	CH_3

The addition of 2,5-dideoxystreptamine or streptamine instead of 2-deoxystreptamine to a deoxystreptamine-requiring mutant of *Bacillus circulans* provides 5-deoxybutirosamine *249* or 2-hydroxybutirosin *250*[267].

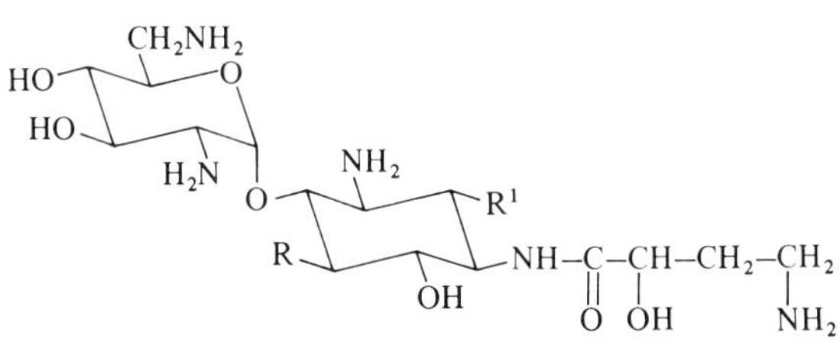

249: R, R^1 = H   *250*: R = Pentosyl, R^1 = OH   **Fig. 56.**

In the same way, from *Micromonospora purpurea* mutants, gentamicin analogues are synthesised with variations in 2- and 5-positions[268, 269].

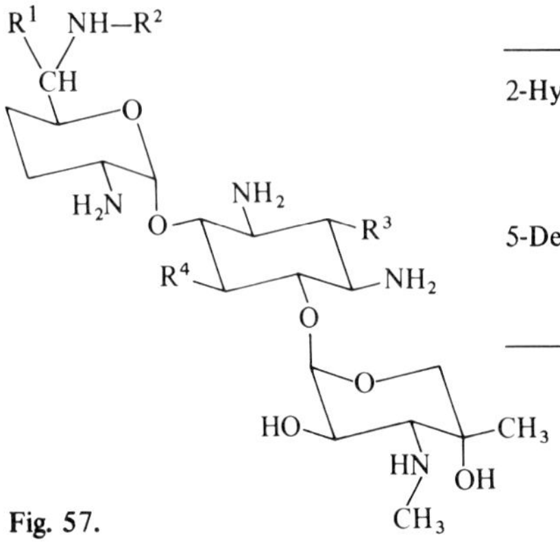

		R^1	R^2	R^3	R^4
2-Hydroxygentamicin	C_1	CH_3	CH_3	OH	OH
	C_2	CH_3	H	OH	OH
	C_{1a}	H	H	OH	OH
5-Deoxygentamicin	C_1	CH_3	CH_3	H	H
	C_2	CH_3	H	H	H
	C_{1a}	H	H	H	H

Fig. 57.

A new antibiotic (streptomutin A) is produced by addition of 2-deoxystreptidine to a streptidine-requiring mutant[263]. The "biotransformation" of 2-deoxystreptamine-containing pseudodi- and trisaccharides into other active aminoglucosides has been reported[265, 270, 271]. The bioconversion of paromamine, neamine and pseudotrisaccharides (gentamicin A, A_2, X_2, antibiotic 66-40B and JI 20A) into sisomicin has been demonstrated, utilising a mutant of *Micromonospora inyoensis,* which requires 2-deoxystreptamine for the sisomicin production[271].

Further examples of biochemical modifications of aminoglycosides are: enzymatic production of 3-N-carboxymethylribostamycin[272], selective microbial hydrolysis of α- or β-glucoside linkages of validamycins[273] and synthesis of aminoglycoside-3′-phosphates[274]. The latter compounds have been used as starting materials for the preparation of 3′-deoxyaminoglycosides via the 3′-chloro derivatives.

IV Structure-Activity Relationships

IV.1 Serum Protein Binding, Complex Formation

In the first instance some physical and chemical properties which influence the efficacy of the aminoglycoside antibiotics will be mentioned. Solubility, stability, lipophilicity, hydrophilicity and chelation properties are important factors for their absorption, transport and bioavailability.

The reversible binding of aminoglycosides to macromolecular structures in blood and tissues is important for biological activity, pharmacokinetics, toxicity and excretion[7, 275, 276]. Binding equilibria of this type are difficult to analyse quantitatively, except for binding to plasma protein, for which several methods of determination (e.g. dialysis, ultrafiltration) exist[277]. Reported figures for serum binding of gentamicin obtained by a variety of methods vary over a wide range. Vogt and Lebek[278)] observed values of about 45%. Black et al.[279)] and Weinstein et al.[280)] found a binding from 25–30%. Later R. C. Gordon et al.[281)], P. Naumann and W. W. Anwärter[282)], and H. Rosenkranz et al.[283)] found no detectable binding of gentamicin to human serum protein over a wide range of concentrations. These large differences are probably caused by experimental errors due to binding of aminoglycosides to membranes, filtration materials and vessels[283, 284)].

Very low protein binding by human serum was found for kanamycin, sisomicin and tobramycin[7, 281–284)]. According to Ramirez-Ronda et al.[284)] the extent of protein binding can change under pathological conditions: in the absence of Ca^{2+}- and Mg^{2+}-ions, aminoglycosides are strongly bound (70–80%) to serum proteins. With normal Ca^{2+}- and Mg^{2+}-levels no binding is observed.

These findings can be explained by the great tendency of the hydroxy- and amino groups to chelate with metal ions. Chelation can be observed by the occurrence of several effects: a drop in pH values, changes of spectra and color, change of solubility and decrease of metal ion concentration in solution[285)].

The first study of aminoglycoside chelates with Cu-, Ni-, Co-, Mg- and Zu ions has been reported by Foye et al.[286)]. They found a 3:1 ratio for the copperchelate, which exhibited more prolonged blood levels in guinea pigs, but also greater toxicity. These chelates are stable at basic pH and are readily dissociated at or below pH 5, where most of the amino groups are protonated. S. Yamabe[287)] noticed the formation of metal chelates with kanamycin. The Cu-chelates with kanamycin and neomycin and their UV-spectra have been studied by A. S. Kershner[36)]. The components of gentamicin C-complex chelate with metal ions (3:2 ratio for the copper complex) with the following order of stability:

$$Cu(II) > Ni(II) > Co(II) > Mn(II) > Mg(II)$$

This order agrees with the results of H. Irving and R. J. P. Williams[288)] for complexes in which the donor atoms are nitrogen and oxygen. Co(III)-complexes, prepared by reaction of gentamicin C_1, C_2 and C_{1a} with $CoCl_2$ in the presence of oxygen, show interesting properties[289)]. Although their pharmacokinetic properties are altered marginally, they display surprisingly the same high activity as the free gentamicins. In an in vivo study more than 80% of the activity was found in the urine after 6 hours. The administered Co(III)-complex was recovered unchanged. The appropriate neomycin-, neamine-, and kanamycin-complexes also possessed good bacteriostatic efficacy.

IV.2 Structure Modifications

Due to the great number of functional groups and asymmetric centres, only a small portion of the possible structural variations has been synthesised and tested until now. The structural variety of antibacterial aminoglycosides makes it difficult to present a comprehensive picture for the structure-activity relationships.

It has been known for a long time that the removal of both the amidino groups in streptomycin and the acylation of the amino groups destroys the antibiotic activity, whereas the reductive amination, the reduction to dihydrostreptomycin, the monosubstitution of an amidino function by a carbonyl group as in bluensomycin and the demethylation of the N-methylamino group do not affect the effectiveness of streptomycin. The differences in the activity of kanamycin A and B and the neomycins have shown that the number and the position of the amino groups are of importance for their efficacy.

Substantial progress towards the improvement of the activity of the aminoglycosides was made by the clarification of the enzymatic inactivation mechanisms of aminoglycoside by resistant organisms. The formation of those enzymes which inactivate aminoglycosides is determined by episomal genes. This type of resistance is therefore called episomal resistance or R-factor mediated resistance, which can be transferred between bacteria cells, but not by spontaneous mutation.

The genetic mechanisms of resistance and especially the R-factor mediated resistance have been surveyed in numerous reviews, for instance by S. Mitsuhashi[10, 290, 291a)], T. Watanabe[291b)], S. N. Cohen[291c)], E. Reinhard[292)] and N. Tanaka[9)].

The studies of the biochemical resistance, i.e. the studies of the enzymatic inactivation steps have been pursued intensively by H. Umezawa[5, 293)] and J. Davies[293–295)] and their co-workers. A summary of their results is compiled in Table 10:

Table 10.

Enzymes	Substrates
3''-O-Phosphotransferase	Streptomycin
3'-O-Phosphotransferase	Neomycin, kanamycin, ribostamycin, butirosin
5''-O-Phosphotransferase	Lividomycin, ribostamycin
2'-N-Acetyltransferase	Gentamicin, tobramycin, lividomycin a.o. not: kanamycin A, amikacin
6'-N-Acetyltransferase	Kanamycin, gentamicin C_1 tobramycin, ribostamycin, sisomicin
3-N-Acetyltransferase	Gentamicin, sisomicin, kanamycin
3''-O-Adenyltransferase	Streptomycin, spectinomycin
2''-O-Nucleotidyltransferase	Gentamicin, kanamycin, tobramycin

The enzymes listed in Table 10 have been discovered in resistant strains of different bacteria. These enzymes catalyze the transfer of a phosphoryl-, acetyl-, or adenyl-moiety to an aminoglycoside to make it inactive.

Due to the knowledge of the enzymatic inactivation process, a great number of chemical transformations has been accomplished with the object of preventing these inactivating reactions. Full details of the chemical modifications have been given in Chap. III.2. These variations are summarised with respect to structureactivity relationships (Ref.[6]).

There are several ways of preventing an enzymatic reaction with an hydroxy-, amino-, or guanidino group:

1. removal of hydroxy- or amino groups
2. establishment of steric hindrance for reduced interaction with the enzyme
3. substitution on the hydroxyl- and amino functions or their exchange for functional groups which cannot be attacked by the enzymes.
4. Replacement of a complete sugar moiety by another carbohydrate with different stereochemistry and/or other substituents.

The removal of the 3′-hydroxyl groups causes in general an enhanced activity against resistant organisms with retention of effectiveness against sensitive strains, (3′-deoxykanamycin A (*169*), 3′-deoxykanamycin B or tobramycin (*30, 174*), 3′-deoxyneamine (*86*) and 3′-deoxybutiroscn B (*111*)).

4′-Deoxykanamycin A (*173*), in contrast to kanamycin A, is not phosphorylated by 3′-O-phosphotransferase II, while the 3′-O-phosphotransferase I producing strains show resistance. Therefore, the 4′-hydroxyl group must be involved in the binding with the enzyme phosphotransferase II. By a study of the effect of the 4′-deoxygenation with neamine, 1-N-HABA-neamine, ribostamycin and butirosin, it could be established that the inactivating reaction of the phosphotransferase II can only be prevented, if the 1-amino function is substituted by an amino acid side chain (HABA).

The 3′,4′-dideoxy compounds show in respect of activity and behavior against resistant strains a spectrum similar to that of the 3′-deoxy derivatives (3′,4′-dideoxyneamine (*87*), 3′,4′-dideoxykanamycin B (*176*)). The advantageous effect of a combination of the deoxygenation and 1-amino-HABA substitution was shown for 3′,4′-dideoxyribostamycin (*105*) and 3′,4′-dideoxybutirosin A (*112*) and B (*113*)[174].

Attempts to prevent the 3′-O-phosphorylation by protection of the hydroxyl groups (3′-O-methylneamine (*92*), 4′-O-methylneamine (*93*) and 3′-O-methylkanamycin A (*170*)) led to inactive compounds. The studies of H. Umezawa et al.[211] indicate that the protection of the 3′-hydroxyl group causes a steric hindrance of the binding of the antibiotic to the bacterial ribosomes. These compounds only marginally inhibit the poly-phenylalanine synthesis in a poly-U-ribosome system.

Phosphotransferase I phosphorylates not only the 3′-hydroxyl group, but also the 5″-hydroxyl function of lividomycin and 3′,4′-dideoxyribostamycin. The exclusive attack at the 3′-position of ribostamycin indicates that this position is sterically better located for the phosphate-transferring center of the enzyme than the 5″-hydroxyl group. A 5″-O-phosphotransferase that reacts also in this position of ribostamycin has been isolated from *Pseudomonas.* The interesting point of these 5″-O-phosphorylations is that the 1-amino-HABA moiety in butirosin obviously weakens the substrate-enzyme binding, so that the 5″-O-phosphorylation does not occur. The removal or the protection of the 5″-hydroxyl function leads to different results. 5″-deoxylividomycin A (*150*) and B (*149*), 5″-deoxy-5″-aminolividomycin A (*151*) and 5″-O-methyllividomycin possess only weak antibacterial activity. Thus the 5″-hydroxyl group seems to be essential for the activity[192]. It is also the reason for the decrease of activity if the 3′,4′-dideoxyribostamycin (*105*) is converted into 3′,4′,5″-tri-deoxyribostamycin (*106*)[171]. For 6′,5″-diamino-6′,5″-dideoxylividomycin B (*153*) and 5″-amino-3′,4′,5″-trideoxybutirosin *A* (*118*)[178, 179] on the other hand, a strong activity against sensitive and resistant organisms has been observed.

Oxidation of ribostamycin to ribostamycin-5″-uronic acid and its derivatives destroys the antibacterial activity.

Modifications of the 6′-amino function, which are believed to prevent a reaction with the 6′-N-acetyltransferase were confirmed mainly so far to alkylation reactions. The antibacterial activity of 6′-N-methylkanamycin B (*177*) and 3′,4′-dideoxy-6′-N-methylkanamycin B (*178*) is comparable with that of the parent compound. They are active against some 6′-N-acetyltransferase-producing bacteria. The 2″-hydroxyl group in kanamycin- or gentamicin-type compounds is attacked by different resistant organisms with adenylation. Thus 2″-deoxygentamicin C_2 (*226*) shows good activity against these organisms but against other gentamycin resistant bacteria it is as inactive as 2″-deoxy-3″-demethylamino-2″ methylaminogentamicin (*227*). The epimer, 2″-epigentamicin C_1 (*225*) is not a substrate for the gentamicin adenyl synthetase. Nevertheless, it shows only a low antibacterial activity[238 a)].

Kanamycin derivatives with a changed configuration in C-2″-position such as 2″-manno-kanamycin A, 6″-amino-6′-deoxy-2″-manno-kanamycin (*161*) and 6″-deoxy-6″-hydrazino-2″-manno-kanamycin A (*162*) have a decreased activity compared with the parent compounds.

The changed stereochemistry obviously prevents not only the interaction with the active center of the inactivating enzyme, but also the reaction with the bacteria ribosomes.

The kanamycin- and gentamicin-type antibiotics need an equatorial 2″-hydroxyl group for a maximal activity against a broad spectrum of organisms. From the model of the kanamycin molecule, the 1-amino group of the 2-deoxystreptamine moiety is seen to be close to the 2″-hydroxyl groups at which the nucleotidyltransferase reacts. Through the modification of the 1-amino function, a steric hindrance of the enzyme reaction can be achieved by which active derivatives have been obtained. Thus, the known 1-amino-HABA substitution inhibits not only the shown attack of the phosphotransferase, but also the reaction of the 2″-hydroxyl nucleotidyltransferase. Extensive studies have been carried out on the structure activity relations of the 1-amino derivatives (HABA and others) of neamine[161], ribostamycin (butirosin analogues)[167] and kanamycin A (amikacin analogues)[232]. It was established unambiguously that the hydroxyl- and amino functions must be present in the acyl side chain to obtain high antibacterial activity. 2,4-diaminobutyrylneamine, 2,3,4,6-tetrahydroxy-5-aminohexanoylribostamycin and 3-aminopropionylkanamycin A, for instance, show no activity. The α-hydroxy-ω-amino acids with 3–5 C-atoms (preferentially 4c atoms) and an S-configuration of the hydroxyl group give optimal effects. The conversion of the ω-amino function into a stronger basic function (guanidino group) causes a loss of activity. Substitution of the 1-amino function with unsubstituted acyl groups or alkyl moieties also decreases the antibacterial activity.

On the other hand, the 1-N-acetyl derivatives of sisomicin and verdamicin possess activity against sensitive organisms comparable with the parent compounds and even enhanced activity against sisomicin-resistant strains carrying 3-N-acetylating and 2″-O-adenylating enzymes[249]. In contrast 1-N-[(S)-α-hydroxy-ω-aminoacyl]-derivatives of sisomicin, verdamicin and gentamicin C_1 are much less effective than, for instance, amikacin. This seems to be a general property of amino glycosides with a 2′-amino group.

1-N-ethylsisomicin (netilmicin) (*241*) possesses excellent activity over a broad spectrum and low toxicity[256a].

Hanessian et al.[296] investigated the structure activity relations of isomeric 2-deoxystreptamine derivatives which were prepared by oxidative degradation of paromomycin (*251*) or via O-glycoside synthesis (*252*).

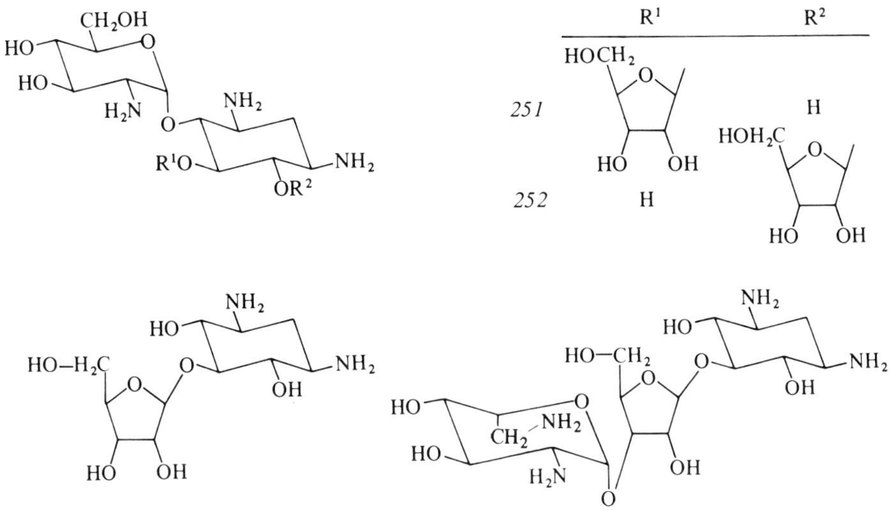

Fig. 58. *253* *254*

The 6-ribofuranosyl derivative is much less active than *251*.

Products of the oxidative degradation (*253, 254*), in which the 4–0 attached aminoglycosyl moiety is removed, show less or no activity[297]. K. Fujisawa et al.[158a] studied the structure-toxicity relations of 25 natural and semi-synthetic aminoglycosides. They evaluated the acute, intravenous toxicity [LD_{50}] with the assumption that the acute toxicity is related to the oto- and nephrotoxicity, which are the limiting side effect in clinical use. The results are summarized in Table 11 [185a]:

Table 11.

Aminoglycoside		LD_{50} in mg/kg/Mouse
8	Neamine	125 (121–129)
34	4'-Deoxyneamine	195 (181–209)
89	3',4',5,6-Tetradeoxyneamine	60 (44– 76)
96	1-N-HABA-Neamine	260 (247–273)
	4'-Deoxy-1-HABA-Neamine	330 (318–342)
10	Neomycin B	24 (21– 27)
11	Neomycin C	44 (37– 51)
12	Paromomycin I	160 (145–175)
15	Lividomycin B	140 (130–150)
14	Lividomycin A	280 (252–308)
18	Ribostamycin	260 (250–270)
19	Xylostasin	280 (265–295)
	4'-Deoxycylostasin	280 (260–300)
20	Butirosin A	520 (502–538)
22	4'-Deoxybutirosin A	520 (515–526)
25	Bu 1709 E_1	890 (656–944)
27	Kanamycin A	280 (269–291)
28	Kanamycin B	132 (124–140)
29	Kanamycin C	225 (198–252)
30	Tobramycin	79 (74– 84)
176	3',4'-Dideoxykanamycin B	71 (67– 75)
37	Gentamicin C_1	88 (78– 98)
38	Gentamicin C_2	70 (65– 75)
39	Gentamicin C_{1a}	70 (59– 81)
179	Amikacin	300 (285–315)

V Mode of Action and Side Effects of Aminoglycoside Antibiotics

Aminoglycoside antibiotics inhibit the protein biosynthesis in sensitive organisms. The reaction takes place at the ribosomes. (Reviews see Refs.[9, 113, 298, 299].) The reaction with the ribosomes causes two effects:

1. Inhibition of the translation step in proteinbiosynthesis.
2. Misreading of the nucleotide code and formation of altered sequences.

As shown for streptomycin, the aminoglycosides bind at the 30 S subunit of the ribosome. The formation of the initiation complex is not inhibited. The translocation of the formylmethionine or peptidyl-t-RNA from the acceptor to the peptidyl site is disturbed. The bound t-RNA shifts the ribosome during the translocation from the decoding site to the condensing site and no more aminoacyl-t-RNA is bound.

Concerning the binding of the agents at the ribosomes and the molecular mechanisms which cause the abovementioned effects, see Refs.[300–304].

Comprehensive studies on the mode of action, in particular of the misreading of aminoglycosides, have been carried out by D. Lando et al.[305]. Corresponding to their different modes of action, the aminoglycosides can be divided into three groups:

1. Kasugamycin produces no misreading, but inhibits the commencement of translation.
2. Streptomycin and dihydrostreptomycin inhibit the initiation step of translation, and there are also slight "misreading" effects.
3. Neomycin, gentamicin, kanamycin and their derivatives produce considerable misreading effects. Neomycin inhibits on a small scale the initiation step of translation, while gentamicin and kanamycin do not show this effect.

Little is known about the molecular mechanism which is responsible for the toxic side effects[306, 307].

For clinical use, the toxic side effects, which appear after long treatment with therapeutic doses, play a more important role than the acute toxicity caused by very high doses (Chap. IV.2).

The ototoxicity causes serious irreversible damage of the hair cells in the inner ear[308]. The area where the first damage is observed is characteristic for the type of the aminoglycoside used and is related to its specific pharmacokinetics. Due to the fact that aminoglycosides are excreted faster from the serum than from the inner ear a cumulative effect in the perilymph is observed. By an exact adjustment of the daily dose and its subdivision into small doses over shorter periods of time according to the specific pharmacokinetics in the patient, toxic levels of the antibiotic in the perilymph can be avoided.

The nephrotoxicity is often reversible but it can potentiate the ototoxicity by a slow excretion of the drug in the impaired kidney.

Many attempts have been made to reduce the toxicity of aminoglycosides. The combination of D-glucaro-δ-lactam decrease the nephrotoxicity[309]. The application of 2,3-dimercaptopropanol during the therapy reduces the ototoxicity.

VI References

1. Schatz, A., Bugie, E., Waksman, S. A.: Proc. Soc. exp. Biol. Med. *55*, 66 (1944)
2a. Schering Co.: U.S. Patent 3.091 572, 14. 11. 1961, b. Weinstein, M. J.: J. Med. Chem. *6*, 463 (1963)
3. Cooper, D. J.: J. Pure and Applied Chemistry *28*, 455 (1971)
4. Umezawa, S.: Adv. Carbohydrate Chem. Biochem. *30*, 111 (1974)
5. Umezawa, H.: Adv. Carbohydrate Chem. Biochem. *30*, 183 (1974)
6. Price, K. E., Godfrey, J. C., Kawaguchi, H.: Adv. Appl. Microbiology *18*, 191 (1974)
6a. Price, K. E., Godfrey, J. C., Kawaguchi, H.: in: Structure-activity relationships among the semisynthetic antibiotics. Edited by D. Perlman. New York: Academic Press, 1977
7. Grassi, G. G.: Future Trends in Chemotherapie Symposium held in Tirrenia (Pisa) 6.–7. May 1974 -Med. Act., Drugs of Today, *10*, Suppl. Nov., 99 (1974)
8. Yagisawa, Y.: Japan Med. Gaz. March 20, 1975
9. Tanaka, N.: Aminoglycoside antibiotics. Antibiotics III: Mechanism of action of antimicrobial and antitumor agents. Corcoran, J. W., Hahn, F. E. (eds). Springer: Berlin, Heidelberg, New York 1975, 340–364
10. Mitsuhashi, S. (ed.): Drug action and drug resistance in bacteria; 2. aminoglycoside antibiotics. University Park Press: Tokyo 1975
10a. After completion of this manuscript a further review has been published: Cox, D. A., Richardson, K., Ross, B. C.: Top. Antibiot. Chem. 1977 *1*, 2–90

11. Heding, H.: Acta chem. Scand. *23,* 1275 (1969)
12. Stodola, F. H., Shotwell, O. L., Borud, A. M., Benedict, R. G.: J. Am. Soc. *73,* 2290 (1951)
13. Fried, J., Titus, E.: J. Biol. Chem. *168,* 391 (1947)
14a. Arcamone, F., Cassinelli, G., D'Amico, G., Orezzi, P.: Experienta *24,* 441 (1968); b) Societa Farmaceutici Italia; Brit. Pat. 955 762, 22. 4. 1964 [C.A. *61,* 2440 (1964)]
15a. Tatsuoka, S., Kusaka, T., Miyake, A., Inoure, M., Hitomi, H., Shiraishi, Y., Iwasaki, H., Imanishi, M.: Pharmac. Bull. (Tokyo) *5,* 343 (1957); b. Kavanagh, F., Grinnan, E., Allanson, E., Tumin, D.: Appl. Microbiol. *8,* 160 (1960)
16. Bannister, B., Argoudelis, A. D.: J. Am. Chem. Soc. *85,* 234 (1963)
17. Peck, R. L., Hoffhine, Jr., C. E., Gale, P. H., Folkers, K.: J. Am. Chem. Soc. *75,* 1018 (1953)
18. Meiji Seika Kaisha Ltd.: JA 7 129 192, 7. 6. 1967
19a. Waksman, S. A., Lechevalier, H. A.: Science *109,* 305 (1949); b. Rinehart, Jr., K. L.: The neomycins and related antibiotics. Squibb, E. R. Lectures on Chemistry of Microbial Products John Wiley and Sons, Inc.: New York 1964
20a. Haskell, T. H., French, J. C., Bartz, O. R.: J. Am. Chem. Soc. *81,* 3482 (1959); b. Parke, Davis Co: Belg. Pat. 547 976 12. 10. 1956
21. Rinehart, Jr., K. L., Hichens, M., Argoudelis, A. D., Chilton, W. S., Carter, H. E., Georgiadis, M., Schaffner, C. P., Schillings, R. T.: J. Am. Chem. Soc. *84,* 3218 (1962)
22a. Mori, T., Ichiyanagi, T., Kondo, H., Tokunaga, K., Oda, T., Munakata, K.: J. Antibiotics *24,* 339 (1971); b. Kowa Co, Ltd.: DAS 1 767 835, 21. 6. 1967
23a. Shier, W. T., Rinehart, Jr., K. L., Gottlieb, D.: Proc. Nat. Acad. Sci. U. S., *63,* 198 (1969); b. Shier, W. T., Schaefer, P. C., Gottlieb, D., Rinehart, K. L.: Biochemistry *13,* 5073 (1974); c. University Illinois Found.: U. S.-Pat. 3 833 556, 17. 7. 1969
24a. Akita, E., Tsuruoka, T., Ezaki, N., Niida, T.: J. Antibiotics *23,* 155, 173 (1970); b. Meiji Seika Kaisha Ltd.: NL 6 818 105, 18. 12. 1967
25. Takeda Chem. Ind. Ltd. Osaka, (Japan): DOS 2 326 943, 31. 5. 1972
26a. Dion, H. W., Woo, P. W. K., Willmer, N. E., Kern, D. L., Onaga, J., Fusari, S. A.: Antimicrobial Agents Chemoth. *2,* 84 (1972); b. Parke Davis Co.: U.S.-Pat. 3 541 078, 15. 5. 1968
27a. Konishi, M., Numata, K., Shimoda, K., Tsukiura, H., Kawaguchi, H.: J. Antibiotics *27,* 471 (1974); b. Bristol Meyers Co.: DOS 2 346 243, 13. 9. 1972
28a.Kirby, J. P., Borders, D. B., Van Lear, G. E.: J. Antibiotics *30,* 175 (1977); b. American Cyanamid Co.: U.S.-Pat. 3 928 317, 25. 2. 1974
29. Tsukiura, H., Saito, K., Kobaru, S., Konishi, M., Kawaguchi, H.: J. Antibiotics *26,* 386 (1973)
30. Umezawa, H.: J. Antibiotics *A 10,* 107 (1957)
31a. Wakazawa, T., Sugano, Y., Abe, M., Fukatsu, S., Kawaji, S.: J. Antibiotics *A 14,* 180 (1961); b. Meiji Seika Kaisha Ltd., Tokyo: DAS 1 793 403, 12. 9. 1968
32. Murase, M., Wakazawa, T., Abe, M., Kawaji, S.: J. Antibiotics, Ser. *A 14,* 156, 367 (1961); a. Toda, S., Nakagawa, S., Naito, T.: J. Antibiotics *30,* 1002 (1977)
33a. Koch, K. F., Rhoades, J. A.: Antimicrobial Agents and Chemother. *1970,* 309 and[10) p. 113–121; b. Koch, K. F., Davis, F. A., Rhoades, J. A.: J. Antibiotics *26,* 745 (1973)
34a. Nara, T., Takasawa, S., Yamamoto, M., Sato, S., Sato, T., Okachi, R., Kawamoto, I.: 15th Interscience Conf. Antimicrobial Agents Chemotherapie, Washington D. C. Sept. 1975, 30; b. Kyowa Fermentation KK: DOS 2 450 411, 24. 10. 1973; Abbott Laboratories; Belg.-Patent 826 936, 20. 3. 1975; c. Nara, T. et al.: J. Antibiotics 17 (1977)
35. Schering Corp.: U.S.-Pat. 3 091 572, 16. 7. 1962 and [2, 3)]
36. Kershner, A. S.: Ph.D.Thesis, Rutgers University, State University of New Jersey 1971
37a. Egan, R. S., Devault, R. L., Mueller, S. L., Levenberg, M. I., Sinclair, A. C., Stanaszek, R. S.: J. Antibiotics *28,* 29 (1975); b. Daniels, P. J. L., Luce, C., Nagabhushan, T. L, Jaret, R. S., Schumacher, D., Reimann, H., Ilavsky, J.: J. Antibiotics *28,* 35 (1975); c. Abbott Laboratories, Kyowa Hakko Kogyo Co. Ltd., DOS 2 326 781, 27. 5. 1972
38. Maehr, H., Schaffner, C. P.: J. Am. Chem. Soc. *92,* 1697 (1970)
39. Nagabhushan, T. L., Turner, W. N., Daniels, P. J. L., Morton, J. B.: J. Org. Chem. *40,* 2830 (1975)
40. Nagabhushan, T. L., Daniels, P. J. L., Jaret, R. S., Morton, J. B.: J. Org. Chem. *40,* 2835 (1975)

41. Schering Corp.: U.S.-Pat. 3 915 955, 22. 6. 1970
42. Scherico Ltd., Luzern (Switzerland): Belg. Pat. 787 758, 8. 11. 1971
43. Scherico Ltd., Luzern (Switzerland): DOS 2 329 012, 12. 6. 1972
44. Bérdy, J., Pauncz, J. K., Vajna, Zs. M., Horvath, Gy., Gyimesi, J., Koczka, I: J. Antibiotics *30,* 945 (1977)
45a. Cooper, D. J., Jaret, R. S., Reimann, H.: Chem. Commun. *1971,* 285; b. Scherico Ltd., Luzern (Switzerland): DAS 1 932 309, 27. 6. 1968
46. Weinstein, M. J., Wagman, G. H., Marquez, J. A., Testa, R. T., Waitz, J. A.: Antimicrobial Agents Chemoth. *7,* 246 (1975)
47a. Daniels, P. J. L., Jaret, R. S., Nagabhushan, T. L., Turner, W. N.: J. Antibiotics *29,* 488 (1976); b. Scherico Ltd., Luzern (Switzerland): DOS 2 334 923, 14. 7. 1972
48a. Davies, D. H., Greeves, D., Mallams, A. K., Morton, J. B., Tkach, R. W.: J. Chem. Soc. Perkin Trans. I, *1975,* 814; b. Scherico Ltd., Luzern (Switzerland): BE 811-370, 23. 2. 1973
49. O'Connor, S., Lam, L. K. T., Jones, N. D., Chaney, M. O.: J. Org. Chem. *41,* 2087 (1976)
50a. Kondo, S., Iinuma, K., Naganawa, H., Shimura, M., Sekizawa, Y.: J. Antibiotics *28,* 79, 83 (1975); b. Meiji Seika Kaisha Ltd.: DAS 1 642 305, 8. 7. 1966
51. Mann, R. L., Bromer, W. W.: J. Am. Chem. Soc. *80,* 2714 (1958)
52. Inouye, S., Shomyra, T., Watanabe, H., Totsugawa, K., Niida, T.: J. Antibiotics *26,* 374 (1973)
53a. Nara, T., Okachi, R., Yamamoto, M., Kawamoto, I., Takayama, K., Takasawa, S., Sato, T., Sato, S.: Interscience Conf. Antimicrobial Agents and Chemotherapy. 1976, Chicago, Abstr. 56–58; b. Abbott Labs.: Belg.-Patent 817 954, 23. 7. 1973; c. Kyowa Hakko Kogyo Co., Ltd. Tokio: DOS 2 418 349, 17. 4. 1973; d. Egan, R. S., Stanaszek, R. S., Cirovic, M., Mueller, S. L., Tadanier, J.: J. Antibiotics *30,* 552 (1977). e. Abbott Laboratories: BE 844-756, 1. 8. 1975
54a. Cochran, T. G., Abraham, D. J., Martin, L. L.: Chem. Commun. *1972,* 494 (1972); b. Hoeksma, H., Knight, J. C.: Antibiotics *28,* 240 (1975)
55. Horii, S., Kameda, Y.: Chem. Commun. *1972,* 747
56. Mann, R. L., Woolf, D. O.: J. Am. Chem. Soc. *79,* 120 (1957)
57. Iinuma, K., Kondo, S., Maeda, K., Umezawa, H.: J. Antibiotics *28,* 613 (1975)
58a. Suhara, Y., Maeda, K., Umezawa, H., Ohno, M.: Tetrahedron Lett. *1966,* 1239; b. Inst. of Microbial Chemistry: DAS 1 617 951, 11. 10. 1966
59. Konishi, M., Kamata, S., Tsuno, T., Numata, K., Tsukiura, H., Naito, T., Kawaguchi, H.: J. Antibiotics *29,* 1152 (1976)
60a. American Cyanamid Co.: U.S.-Pat. 3 987-029, 23. 1. 1974; b. Kirby, J. P., Van Lear, G. E., Morton, G. O., Gore, W. E., Curran, W. V., Borders, D. B.: J. Antibiotics *30,* 344 (1977)
61. Mathis, C.: Bull. Soc. Chim. France *1973,* 93
62. Pauncz, J. K.: J. Antibiotics *25,* 677 (1972)
63. Omoto, S., Inouye, S., Niida, T.: J. Antibiotics *24,* 430 (1971)
64. Wagman, G. H., Weinstein, M. J.: Chromatogr. Lib. Vol. 1, Chromatography of antibiotics. Amsterdam: Elsevier Publishing Co. 1973.
65. Warren, E., Snyder, R. J., Washington, J. A.: Antimicrobial Agents Chemother. *1,* 46 (1972)
66. Lamb, J. W., Mann, J. M., Simmons, R. J.: Antimicrobial Agents Chemother. *1,* 323 (1972)
67. Lund, M. E., Blazevic, D. J., Matsen, J. M.: Antimicrobial Agents Chemother. *4,* 569 (1973)
68. Acid, D. V., Seligman, S. J.: Antimicrobial Agents Chemother. *3,* 559 (1973)
69. Marengo, P. B., Wilkins, J., Overturf, G. D.: Antimicrobial Agents Chemother. *6,* 498 (1974)
70. Haas, M. J., Davies, J.: Antimicrobial Agents Chemother. *4,* 497 (1973)
71. Broughall, J. M., Reeves, D. S.: J. Clin. Pathol. *28,* 140 (1975)
72. Stevens, P., Young, L. S., Hewitt, W. L.: Antimicrobial Agents Chemother. *7,* 374 (1975)
73. Smith, D. H., Otto B. V., Smith, A. L.: New Engl. J. Med. *286,* 583 (1972)
74. Smith, A. L., Smith, D. H.: J. Infectious Dis. *129,* 391 (1974)
75. Holmes, R. K., Sanford, J. P.: J. Infectious Dis. *129,* 519 (1974)
76 Mahon, W. H., Ezer, J., Wilson, T. W.: Antimicrobial Agents Chemother. *3,* 585 (1973)
77. Lewis, J. E., Nelson, J. C., Elder, H. A.: Nature (New Biol.) *239,* 214 (1972)
78. Lewis, J. E., Nelson, J. C., Elder, H. A.: Antimicrobial Agents Chemother. *7,* 42 (1975)

79. Stevens, P., Young, L. S., Hewitt, W. L.: J. Lab. Clin. Med. *86,* 349 (1975)
80. Naganawa, H., Kondo, S., Maeda, K., Umezawa, H.: J. Antibiotics *24,* 823 (1971)
81. Hasegawa, A., Nishimura, D., Nakajima, M.: Agr. Biol. Chem. *36,* 1043 (1972)
82. Inouye, S.: Chem. Pharm. Bull. *14,* 1210 (1966)
83. Kotowycz, G., Lemieux, R. U.: Chem. Rev. *73,* 669 (1973)
84. Reeves, R. E.: Adv. Carbohyd. Chem. *6,* 107 (1951)
85. Neidle, S., Rogers, D., Hursthouse, M. B.: Tetrahedron Lett. 4725 (1968)
86. Bukhari, S. T. K., Guthrie, R. D., Scott, A. I., Wrixon, A. D.: Tetrahedron *26,* 3653 (1970)
87. Morton, J. B., Long, R. C., Daniels, P. J. L., Tkach, R. W., Goldstein, J. H.: J. Am. Chem. Soc. *95,* 7464 (1973)
88. Omoto, S., Inouye, S., Kojima, M., Niida, T.: J. Antibiotics (Tokio) *26,* 717 (1973)
89. Yamaoka, N., Usui, T., Sugiyama, H., Seto, S.: Chem. Pharm. Bull. *22,* 2196 (1974)
90. Koch, K. F., Rhoades, J. A., Hagaman, E. W., Wenkert, E.: J. Am. Chem. Soc. *96,* 3300 (1974)
91. Bock, K., Pedersen, C., Heding, H.: J. Antibiotics *27,* 139 (1974)
92. Nagabhushan, T. L., Daniels, P. J. L.: Tetrahedron Lett. *1975,* 747
93. Inouye, S.: Chem. Pharm. Bull. *20,* 2331 (1972)
94. Dejongh, D. C., Hills, E. B., Hribar, J. D., Hanessian, S., Chang, T.: Tetrahedron *29,* 2707 (1973)
95. Daniels, P. J. L., Mallams, A. K., Weinstein, J., Wright, J. J., Milne, G. W. A.: J. Chem. Soc. *1976,* 1078
96. Rinehart, Jr., K. L., Carter, Jr., J. C., Maurer, K. H., Rapp, U.: J. Antibiotics *27,* 1 (1974)
97a. Summary: see Ref.[7]; b. Walter, A. M., Heilmeyer, L.: Antibiotika-Fibel, 3. Aufl. Stuttgart: Thieme 1969
98. Chemother. Agents in: Annual reports in medicinal chemistry. New York: Academic Press. a. *1974,* 98; b. *1973,* 104; c. *1972,* 99
99. Gilbert, D. N., Kutscher, E., Ireland, P., Barnett, J. A., Sanford, J. P.: J. Infect. Diseases, *124,* Suppl. p. 37 (1971)
100. Davies, S. D., Iametta, A.: Chemotherapy *19,* 243 (1973)
101. Crawford, L. M.: Am. J. Vet. Res. *33,* 1685 (1972)
102. Dienstag, J., Neu, H. C.: Antimicrob. Agents Chemother. *1,* 41 (1972)
103. Denis, F., Geslin, M., Boilleau, Y.: Compt. Rend. Soc. Biol. 164, Nr. 10, 2093 (1970)
104. D'Amato, R. F., Thornsberry, E., Baker, C. N., Kirren, L. A.: Antimicrob. Ag. Chemother. *7,* 596 (1975)
105. Ramirez-Ronda, C. H., Holmes, R. K., Sanford, J. P.: Antimicrob. Ag. Chemother. *7,* 239 (1975)
106. Zimelis, V. M., Jackson, G. G.: J. Infect. Dis. *127,* Nr. 6, 663 (1973)
107. Brogden, R. N., Pinder, R. M., Sawyer, P. R., Speight, T. M., Avery, G. S.: Drugs *12,* 166 (1976)
108. Kobayashi, F., Nagoya, T., Yoshimura, Y., Kaneko, K., Ogata, S., Goto, S.: J. Antibiotics *25,* 128 (1972)
109. Heifetz, C. L., Fisher, M. W., Chodubski, J. A., DeCarlo, M. O.: Antimicrob. Ag. Chemother. *2,* 89 (1972)
110. Klastersky, J., Hensges, C., Gerard, M., Daneau, D.: J. Clin. Pharmacol. *15,* 252 (1975)
111. Walter, A. M., Heilmeyer, L.: see Ref.[96b]
112. Zimmermann, R. A., Moellering, Jr., R. C., Weinberg, A. N.: J. Bacteriol. *105,* 873 (1971)
113. Schlesinger, D., Medoff, G.: see Ref.[9] pg. 535–550
114. Wulff, G., Röhle, G.: Angew. Chem. *86,* 173 (1974)
115. Koenigs, W., Knorr, E.: Ber. dtsch. chem. Ges. *34,* 957 (1901), see also Ref.[113]
116a. Lemieux, R. U., Nagabhushan, T. L., O'Neill, I. K.: Tetrahedron Letters *1964,* 1909; b. Lemieux, R. U., James, K., Nagabhushan, T. L.: Canad. J. Chem. *51,* 42 (1973)
117a. Suhara, Y., Sasaki, F., Maeda, K., Umezawa, H., Ohno, M.: J. Am. Chem. Soc. *90,* 6559 (1968); b. Suhara, Y., Sasaki, F., Koyama, G., Maeda, K., Umezawa, H., Ohno, M.: J. Am. Chem. Soc. *94,* 6501 (1972); c. Nakajima, M., Shibata, H., Kitahara, K., Takahashi, S., Hasegawa, A.: Tetrahedron Lett. *1968,* 2271

118. Nishimura, Y., Tsuchiya, T., Umezawa, S.: Bull. Chem. Soc. Jap. *43,* 2960 (1970)
119. Lemieux, R. U., Nagabhushan, T. L., Clemetson, K. J., Tucker, L. C. N.: Can. J. Chem. *51,* 53 (1973)
120. Umezawa, S., Tatsuta, K., Tsuchiya, T., Kitazawa, E.: J. Antibiotics *A 20,* 53 (1967)
121. Tatsuta, K., Kitazawa, K., Umezawa, S.: Bull. Chem. Soc. Jap. *40,* 2371 (1967)
122. Koto, S., Tatsuta, K., Kitazawa, E., Umezawa, S.: Bull. Chem. Soc. Jap. *41,* 2769 (1968)
123. Hasegawa, A., Kurihara, N., Nishimura, D., Nakajima, M.: Agr. Biol. Chem. (Tokio) *32,* 1123 (1968)
124. Umezawa, S., Koto, S.: J. Antibiotics (Tokio) *A 19,* 88 (1966); Bull. Chem. Soc. Jap. *39,* 2014 (1966)
125. Umezawa, S., Miyazawa, T., Tsuchiya, T.: J. Antibiotics *25,* 530 (1972)
126. Nishimura, Y., Tsuchiya, T., Umezawa, S.: Bull. Chem. Soc. Jap. *44,* 2521 (1971)
127. Umezawa, S., Tatsuta, K., Koto, S.: J. Antibiotics *21,* 367 (1968); Bull. Chem. Soc. Jap. *42,* 533 (1969)
128a. Nakajima, M., Hasegawa, A., Kurihara, N., Shibata, H., Ueno, T., Nishimura, D.: Tetrahedron Lett. *1968,* 623; b. Hasegawa, A., Kurihara, N., Nishimura, D., Nakajima, M.: Agr. Biol. Chem. (Tokio), *32,* 1130 (1968)
129. Umezawa, S., Koto, S., Tatsuta, K., Hineno, H., Nishimura, Y., Tsumura, T.: a. J. Antibiotics *21,* 462 (1968); b. Bull. Chem. Soc. Jap. *42,* 537 (1969)
130. Umezawa, S., Koto, S., Tatsuta, K., Tsumura, T.: a. J. Antibiotics *21,* 162 (1968); b. Bull. Chem. Soc. Jap. *42,* 529 (1969)
131. Takagi, Y., Miyake, T., Tsuchiya, T., Umezawa, S., Umezawa, H.: J. Antibiotics *26,* 403 (1973)
132. Ito, T., Akita, E., Tsuruoka, T., Niida, T.: a. Agr. Biol. Chem. (Tokio) *34,* 980 (1970); b. Antimicrob. Agents Chemother. *1970,* 33; NL-Pat. Meiji Seika Kaisha Ltd. 7 101 319 (2. 2. 1970)
133. Fukami, H., Kitahara, K. K., Nakajima, M.: Tetrahedron Lett. *1976,* 545
134. Ikeda, D., Tsuchiya, T., Umezawa, S., Umezawa, H.: J. Antibiotics *25,* 741 (1972)
135. Chazan, J. B., Gasc, J. C.: Tetrahedron Lett. *1976,* 3145
136. Sitrin, R. D., Cooper, D. J., Weisbach, J. A.: J. Antibiotics *30,* 836 (1977)
137. Nishimura, Y., Tsuchiya, T., Umezawa, S.: Bull. Chem. Soc. Jap. *44,* 2521 (1971)
138. Lemieux, R. U., Nagabhushan, T. L., Clemetson, K. L., Tucker, L. C. N.: Can. J. Chem. *51,* 53 (1973)
139. Umezawa, S., Nishimura, Y.: J. Antibiotics *30,* 189 (1977)
140a. Umezawa, S., Tsuchiya, T., Yamazaki, T., Sano, H., Takahashi,. Y.: J. Am. Chem. Soc. *96,* 920 (1974); b. Umezawa, S., Yamasaki, T., Kubota, Y., Tsuchiya, T.: Bull. Chem. Soc. Jap. *48,* 563 (1975); c. Umezawa, S., Takahashi, Y., Usui, T., Tsuchiya, T.: J. Antibiotics *27,* 997 (1974)
141. Sano, H., Tsuchiya, T., Kobayashi, S., Hamada, M., Umezawa, S., Umezawa, H.: J. Antibiotics *29,* 978 (1976)
142. Bissett, F. H., Evans, M. E., Parrish, F. W.: Carbohyd. Res. *5,* 184 (1967)
143. Jikihara, T., Tsuchiya, T., Umezawa, S., Umezawa, H.: Bull. Chem. Soc. Jap. *46,* 3507 (1973)
144. Miyai, K., Gros, P. H.: J. Org. Chem. *34,* 1638 (1969)
145. Umezawa, S., Takagi, Y., Tsuchiya, T.: Bull. Chem. Soc. Jap. *44,* 1411 (1971)
146. Ikeda, D., Tsuchiya, T., Umezawa, S., Umezawa, H.: a. J. Antibiotics *25,* 741 (1972); b. Bull. Chem. Soc. Jap. *47,* 3136 (1974)
147. Takeda Chem. Ind. Ltd., Osaka: DOS 2 432 644, 12. 7. 1973
148a. Tipson, R. S.: Advan. Carbohydrate Chem. *8,* 107 (1953); b. Tipson, R. S., Cohen, A.: Carbohydrate Res. *1,* 338 (1965); c. Saeki, H., Takeda, N., Shimada, Y., Ohki, E.: Chem. Pharm. Bull. *24,* 724 (1976)
149. Ohrui, H., Emoto, S.: Agr. Biol. Chem. *32,* 1371 (1968)
150. Hayashi, T., Takeda, N., Saeki, H., Ohki, E.: Chem. Pharm. Bull. (Jap.) *25,* 2134 (1977)
151a. Heding, H., Fredericks, G. J., Lützen, O.: Acta Chem. Scand. *26,* 3251 (1972); b. Heding, H., Lützen, O.: J. Antibiotics *25,* 287 (1972)
152. Schmidt, H., Sych, F. J., Gosda, W.: D. Pharmazie *23,* (4), 161 (1968)

153. Zaidan Hojin Biseibutsu Kagaku: DOS 2 361 159, 8. 12. 72
154. Meiji Seika Kaisha: DOS 2 533 985, 1. 8. 1974
155. Meiji Seika Kaisha: DOS 2 555 479, 11. 12. 1974
156. Smith Kline Corp.: US 3 893 997, 1. 7. 1974
157. Smith Kline Corp.: DOS 2 548 246, 29. 10. 1974
158a. Fujisawa, K., Hoshiya, T., Kawaguchi, H.: J. Antibiotics *27,* 677 (1974); b. Smith Kline Corp.: U.S. 3 963 695, 19. 12. 1974; c. Canas-Rodriguez, A., Galan Ruiz-Poveda, S.: Carbohydrate Res. *58,* 379 (1977)
159. Tsukiura, H., Fujisawa, K., Konishi, M., Saito, K., Numata, K., Ishikawa, H., Miyaki, T., Tomita, K., Kawaguchi, H.: J. Antibiotics *26,* 351 (1973)
160. Parke Davis Co.: US-Patent 3 743 634, 14. 5. 71
161. Saeki, H., Shimada, Y., Ohashi, K., Sugawara, S., Ohki, E.: Chem. Pharm. Bull. *22,* 1151 (1974)
162a. Akita, E., Horiuchi, Y., Yasuda, S.: J. Antibiotics *26,* 365 (1973); b. DOS 2 342 946 25. 8. 1972; c. Kurisu, T., Yamashita, M., Nishimura, Y., Miyake, T., Tsuchiya, T., Umezawa, S.: Bull. Chem. Soc. Jap. *49,* 285 (1976)
163. Zaidan Hojin Biseibutsu Kagaku; DOS 2 436 694, 1. 8. 1973
164. Zaidan Hojin Biseibutsu Kagaku; DOS 2 350 203, 6. 10. 1972
165a. Umezawa, S., Ikeda, D., Tsuchiya, T., Umezawa, H.: J. Antibiotics *26,* 304 (1973); b. Takagi, Y., Ikeda, D., Tsuchiya, T., Umezawa, S. + H.: Bull. Chem. Soc. Jap. *47,* 3139 (1974)
166. Zaidan Hojin Biseibutsu Kagaku; DOS 2 350 169, 6. 10. 1972
167. Haskell, T. H., Rodebaugh, R., Plessas, N., Watson, D., Westland, R. D.: Carbohydrate Res. *28,* 263 (1973)
168a. Ikeda, D., Nagaki, F., Umezawa, S., Tsuchiya, T., Umezawa, H.: J. Antibiotics *28,* 616 (1975); Bull. Chem. Soc. Jap. *49,* 3666 (1976); b. Zaidan Hojin Biseibutsu Kagaku; Belg.-Pat. 841 095, 24. 4. 1975
169. Ikeda, D., Tsuchiya, T., Umezawa, S., Umezawa, H.: J. Antibiotics *26,* 799 (1973)
170. Takeda Chemical Ind.: JA 0029-530, 12. 7. 1973 (Derwent 48460W/29)
171. Ikeda, D., Suzuki, T., Tsuchiya, T., Umezawa, S., Umezawa, H.: Bull. Chem. Soc. Jap. *46,* 3210 (1973)
172. Bristol-Meyers Co.: Belg.-Pat. 835 921, 29. 11. 1974
173a. Saeki, H., Shimada, Y., Ohashi, Y., Tajima, M., Sugawara, S., Ohki, E.: Chem. Pharm. Bull. *22,* 1145 (1974); b. Sankyo Co. Ltd., Tokio: DOS 2 416 917, 11. 4. 1973
174. Ikeda, D., Tsuchiya, T., Umezawa, S., Umezawa, H., Hamada, M.: J. Antibiotics *26,* 307 (1973)
175. Parke Davis Co.: US-Pat. 3970-643, 17. 4. 1975
176a. Culbertson, T. P., Watson, D. R., Haskell, T. H.: J. Antibiotics *26,* 790 (1973); b. Parke Davis Co.: DOS 2 301 540, 13. 1. 1972
177. Bristol Meyers: Belg.-Pat. 816 374, 15. 6. 1973; Parke Davis Co., US-Pat. 3 983-102, 17. 4. 1975
178a. Woo, P. W. K.: J. Antibiotics *28,* 522 (1975); b. Parke Davies Co. Belg.-Pat. 833 132, 6. 9. 1974
179a. Saeki, H., Shimada, Y., Ohki, E., Sugawara, S.: J. Antibiotics *28,* 530 (1975); b. Sankyo Co. Ltd.: Belg.-Pat. 821 440, 25. 10. 1973
180. Yamaguchi, T., Kamiya, K., Mori, T., Oda, T.: J. Antibiotics *30,* 332 (1977)
181. Rech. Ind. Therapeutiques; Franz.-Pat. 1361 393, 29. 3. 1963
182. Rinhart, K. L., Argoudelis, A. D., Goss, W. A., Sohler, A., Schaffner, C. P.: J. Am. Chem. Soc. *82,* 3938 (1960)
183. Wagman, G. H., Weinstein, M. J.: J. Med. Chem. *7,* 800 (1964)
184. Umezawa, S., Ito, Y., Fukatsu, S., Umezawa, H.: J. Antibiotics *12,* 114 (1954)
185a. Boissier, J. R., Philippe, J., Zuckerkandel, F., Ores, B., Teillon, J., Dumont, C., Boilot, Y.: C. R. Acad. Sci. *249,* 1415 (1959); b. Boissier, J. R., Philippe, J., Destombes, P., Dumont, C., Boilot, Y., Braun, M. M., Carfantan, C.: Toxicol. Appl. Pharmacol. *7,* 190 (1965)
186. Pénasse, L., Barthélémy, P., Nominé, G.: Bull. Soc. chim. (France) *1969,* 2391

187. Shier, W. T., Rinehart, Jr., K. L.: J. Antibiotics *26,* 547 (1973)
188. Taniyama, H., Sawada, Y., Tanaka, S.: Chem. Pharm. Bull. *21,* 609 (1973)
189. Soc. Farmaceutici Italia: DOS 2 515-629, 11. 4. 1974
190a. Watanabe, I., Tsuchiya, T., Umezawa, S., Umezawa, H., J. Antibiotics *26,* 802 (1975); b. Microbiochem. Res. Found. DOS 2 411 504, 13. 3. 1973
191. Soc. Farmaceutici Italia: DOS 2 515-602, 29. 4. 1974
192. Umezawa, S., Watanabe, I., Tsuchiya, T., Umezawa, H., Hamada, M.: J. Antibiotics *25,* 617 (1972)
193. Yamamoto, H., Kondo, S., Maeda, K., Umezawa, H.: J. Antibiotics *25,* 487 (1972)
194. Hanessian, S., Massé, R., Capmeau, M. L.: J. Antibiotics *30,* 893 (1977)
195a. Watanabe, I., Tsuchiya, T., Umezawa, S. Umezawa, H.: Bull. Chem. Soc. Jap. 48, *2124* (1975); b. Bristol Myers: US 3 896-106, 4. 6. 1973
196. Microbiochem. Res. Found.: JA 9101-355, 6. 2. 1973
197. Watanabe, I., Ejima, A., Tsuchiya, T., Umezawa, S., Umezawa, H.: Bull. Chem. Soc. Jap. *48,* 2303 (1975)
198. Bristol Myers Co., New York: DOS 2 360 946, 6. 12. 1972
199. Bristol Myers Co., New York: DOS 2 322 576, 4. 5. 1972
200. Kondo, S., Yamamoto, H., Iinuma, K., Maeda, K., Umezawa, H.: J. Antibiotics *29,* 1134 (1976)
201. Cron, M. J., Johnson, D. L., Palermiti, F. M., Perron, Y., Taylor, H. D., Whitehead, D. F., Hooper, I. R.: J. Am. Chem. Soc. *80,* 752 (1958)
202a. Fujii, A., Maeda, K., Umezawa, H.: J. Antibiotics *21,* 340 (1968); b. Zaidan Hojin Biseibutsu Kagaku: JA 7030815, 23. 8. 1967
203. Tsuchiya, T., Umezawa, S.: Bull. Chem. Soc. Jap. *38,* 1181 (1965)
204. Inouye, S.: J. Antibiotics *20,* 6 (1967)
205. Inouye, S.: Chem. Pharm. Bull. *15,* 1888 (1967)
206. Umezawa, S., Tatsuta, K., Tsuchiya, T., Yamamoto, E.: Bull. Chem. Soc. Jap. *40,* 1972 (1967)
207. Kobayashi, T., Tsuchiya, T., Tatsuta, K., Umezawa, S.: J. Antibiotics *23,* 225 (1970)
208. Tatsuta, K., Kobayashi, K., Umezawa, S.: J. Antibiotics *20,* 267 (1967)
209. Inouye, S.: Chem. Pharm. Bull. *16,* 573 (1968)
210a. Umezawa, S., Tsuchiya, T., Muto, R., Nishimura, Y., Umezawa, H.: J. Antibiotics *24,* 274 (1971); b. Umezawa, S., Nishimura, Y., Hineno, H., Watanabe, K., Koike, S., Tsuchiya, T., Umezawa, H.: Bull. Chem. Soc. Jap. *45,* 2847 (1972); c. Zaidan Hojin Biseibutsu Kagaku: DOS 2 161 527, 3. 2. 1971
211. Umezawa, H., Tsuchiya, T., Muto, R., Umezawa, S.: Bull. Chem. Soc. Jap. *45,* 2842 (1972)
212. Takagi, Y., Miyake, T., Tsuchiya, T., Umezawa, S., Umezawa, H.: J. Antibiotics *26,* 403 (1973); Bull. Chem. Soc. (Jap.) *49,* 3649 (1976)
213. Bristol Myers Co.: Belg.-Pat. 835-328, 13. 11. 1974
214. Umezawa, S., Nishimura, Y., Hata, Y., Tsuchiya, T., Yagisawa, M., Umezawa, H.: J. Antibiotics *27,* 722 (1974)
215a. Naito, T., Nakagawa, S., Abe, Y., Fujisawa, K., Kawaguchi, H.: J. Antibiotics *27,* 838 (1974); b. Bristol Myers Co.: US-Pat. 3 886-138, 21. 12. 1973; c. Abe, Y., Nakagawa, S., Fujisawa, K., Naito, T., Kawaguchi, H.: J. Antibiotics *30,* 1004 (1977)
216a. Umezawa, H., Umezawa, S., Tsuchiya, T., Okazaki, Y.: J. Antibiotics *24,* 485 (1971); b. Umezawa, S., Umezawa, H., Okazaki, Y., Tsuchiya, T.: Bull. Chem. Soc. Jap. *45,* 3624 (1972); c. Zaidan Hojin Biseibutsu Kagaku: DOS 2 135 191, 29. 7. 1970; d. Nishimura, T., Tsuchiya, T., Umezawa, S., Umezawa, H.: Bull. Chem. Soc. Jap. *50,* 1580 (1977)
217a. Umezawa, H., Nishimura, Y., Tsuchiya, T., Umezawa, S.: J. Antibiotics *25,* 743 (1972); b. Microbiochem. Res. Found.: JAP 9041-345, 23. 8. 1972
218. Naito, T., Nakagawa, S., Abe, Y., Toda, S., Fujisawa, K., Miyaki, T., Koshiyama, H., Ohkuma, H., Kawaguchi, H.: J. Antibiotics *26,* 297 (1973)
219a. Kawaguchi, H., Naito, T., Nakagawa, S., Fujisawa, K.: J. Antibiotics *25,* 695 (1972); b. Bristol Myers Co., New York: DOS 2 234 315, 13. 7. 1971; c. Bristol Myers Co: NL.-Pat. 7401 517, 7. 2. 1973

220a. Bristol-Banyu Res.: JA 0077-345 14. 11. 1974 (Derwent 64761w/39); b. Fujino, M., Kobayashi, S., Obayashi, M., Fukuda, M., Shinagawa, T., Nishimura, O.: Chem. Pharm. Bull. *22,* 1857 (1974)
221. Sato, H., Kusumi, T., Imaye, K., Kakisawa, H.: Chem. Lett. *1975,* 965; Yamada, Y., Okada, H.: Agr. Biol. Chem. *40,* 1437 (1976)
222. Bristol-Myers Co.: DOS 2 311 524, 8. 3. 1972; Zaidan Hojin Biseibutsu Kagaku: DOS 2 618 009, 24. 4. 1975
223. Kondo, S., Iinuma, K., Yamamoto, H., Maeda, K., Umezawa, H.: J. Antibiotics *26,* 412 (1973)
224. Kondo, S., Iinuma, K., Yamamoto, H., Ikeda, Y., Maeda, K., Umezawa, H.: J. Antibiotics *26,* 705 (1973) and Zaidan Hojin Biseibutsu Kagaku: DOS 2 440 956, 29. 8. 1973
225a. Kondo, S., Iinuma, K., Hamada, M., Maeda, K., Umezawa, H.: J. Antibiotics *27,* 90 (1974); b. Zaidan Hojin Biseibutsu Kagaku: DOS 2 423 591, 15. 5. 1973
226. Bristol-Myers Co.: DOS 2 408 666, 23. 2. 1973
227. Umezawa, H., Iinuma, K., Kondo, S., Maeda, K.: J. Antibiotics *28,* 483 (1975)
228a. Umezawa, H., Iinuma, K., Kondo, S., Hamada, M., Maeda, K.: J. Antibiotics *28,* 340 (1975); b. Bristol-Myers Co.: Belg.-Pat. 834 236, 8. 10. 1974
229. Bristol-Myers Co.: DOS 2 555 405, 9. 12. 1974
230. Pfizer Corp.: DOS 2 547 738, 26. 10. 1974; Richardson, K., Jevons, S., Moore, J. W., Ross, B. C., Wright, J. R.: J. Antibiotics *30,* 843 (1977)
231a. Mori, T., Kyotani, Y., Watanabe, I., Oda, T.: J. Antibiotics *25,* 149 (1972); b. Kowa Co. Ltd.: JAP 4844 234, 12. 10. 1971
232. Naito, T., Nakagawa, S., Narita, Y., Toda, S., Abe, Y., Oka, M., Yamashita, H., Yamasaki, T., Fujisawa, K., Kawaguchi, H.: J. Antibiotics *27,* 851 (1974)
233. Cooper, D. J., Weinstein, J., Waitz, J. A.: J. Med. Chem. *14,* 1118 (1971)
234. Scherico Ltd., Luzern (Switzerland): NL.-Pat. 7 012 428, 25. 8. 1969
235. Pfizer Corp.: Belg.-Pat. 815 910, 5. 6. 1973
236. Merck Co., Inc., Rahway (USA): DOS 2 234 804, 16. 7. 1971
237. Mallams, A. K., Vernay, M. F., Crowe, D. F., Detre, G., Tanabe, M., Yasuda, D. M.: J. Antibiotics *26,* 782 (1973)
238a. Daniels, P. J. L., Weinstein, J., Tkach, R. W., Morton, J.: J. Antibiotics *27,* 150 (1974); b. Schering Corp.: US-Pat. 3 868 360, 24. 10. 1972
239. Schering Corp.: US-Pat. 3 920 628, 24. 10. 1972
240. Nagabhushan, T. L., Daniels, P. J. L.: J. Med. Chem. *17,* 1030 (1974)
241. Bristol Myers Co., New York: DOS 2 332 485, 26. 6. 1972
242. Bristol Myers Co., New York: DOS 2 352 361, 18. 10. 1972
243. Bristol Myers Co., New York: DOS 2 355 348, 6. 11. 1972
244a. Kyowa Fermentation: DOS 2 502 935, 24. 1. 1974; b. Abbott Laboratories: NL-Pat. 7 416-210, 12. 12. 1973
245. Kyowa: DOS 2 509 885, 7. 3. 1974
246. Abbott Laboratories: NL-Pat. 7 416-211, 12. 12. 1973
247. Scherico Ltd., Luzern (Switzerland): NL-Pat. 7 500-690, 19. 3. 1974
248. Daniels, P. J. L., Weinstein, J., Nagabhushan, T. L.: J. Antibiotics *27,* 889 (1974)
249. Wright, J. J., Cooper, A., Daniels, P. J. L., Nagabhushan, T. L., Rane, D., Turner, W. N., Weinstein, J.: J. Antibiotics *29,* 714 (1976)
250. Kyowa Hakko Kogyo Co. Ltd., Tokio: DOS 2 601 490, 21. 1. 1975
251. Kyowa Hakko Kogyo: DOS 2 550 168, 9. 11. 1974
252. Kyowa Fermentation KK: JAP 0126-639, 25. 3. 1974
253. Scherico Ltd.: Belg.-Pat. 835 898, 29. 11. 1974
254. Matsushima, H., Kitaura, K., Mori, Y.: Bull. Chem. Soc. Jap. *50,* 3039 (1977)
255. Matsushima, H., Mori, Y., Kitaura, K.: J. Antibiotics *30,* 890 (1977)
256a. Wright, J. J.: J. Chem. Soc. Chem. Commun. *1976,* 206; b. Scherico Ltd, (Switzerland): Belg.-Pat. 818 431, 6. 8. 1973
257. Daniels, P. J. L.: see Ref.[10], pg. 101
258. Rinehart, Jr., K. L., Stroshane, R. M.: J. Antibiotics *29,* 319 (1976)

259. Basak, K., Majumdar, S. K.: Antimicrob. Agents Chemother. *4,* 6 (1973)
260. Majumdar, M. K., Majumdar, S. K.: a. J. Antibiotics *22,* 174 (1969); b. Biochem. J. *120,* 271 (1970); c. Biochem. J. *122,* 397 (1971)
261. Bandyophadhyay, S. K., Majumdar, S. K.: Antimicrob. Agents Chemother. *5,* 431 (1974)
262a. Lee, B. K., Condon, R. G., Wagman, G. H., Katz, E.: Antimicrob. Agents Chemother. *9,* 151 (1976); b. Lee, B. K., Testa, R. T., Wagman, G. H., Liu, C. M., McDaniel, L., Schaffner, C.: J. Antibiotics *26,* 728 (1973)
263. Nagaoka, K., Demain, A. L.: J. Antibiotics *28,* 627 (1975)
264. Shier, W. Th., Ogawa, S., Hichens, M., Rinehart, K. L.: J. Antibiotics *26,* 551 (1973)
265. Kojima, M., Satoh, A.: J. Antibiotics *26,* 784 (1973)
266a. Testa, R. T., Wagman, G. H., Daniels, P. J. L., Weinstein, M. J.: J. Antibiotics *27,* 917 (1974); b. Scherico Ltd., Luzern (Switzerland): Belg.-Pat. 818 429, 6. 8. 1973; c. Schering Corp., DOS 2 437 159, 20. 2. 1975
267. Taylor, H. D., Schmitz, H.: J. Antibiotics *29,* 532 (1976)
268. Sterling Drug Inc.: DOS 2 606 517, 18. 2. 1975
269. Rosi, D. Drozd, M. L., Goss, W. A., Daum, S. J.: 16. Interscience Conf. Antimicrobial Agents Chemotherapie, Chicago; a. Rosi, D., Goss, W. A., Daum, S. J.: J. Antibiotics *30,* 88 (1977); b. Daum, S. J., Rosi, D., Goss, W. A.: J. Antibiotics *30,* 98 (1977)
270. Claridge, C. A., Bush, J. A., Defuria, M. D., Price, K. E.: Devel. Ind. Microbiol. *15,* 101 (1974)
271. Testa, R. T., Tilley, B. C.: J. Antibiotics *28,* 140, *573* (1975)
272. Kojima, M., Inouye, S., Niida, T.: J. Antibiotics *28,* 48 (1975)
273. Kameda, Y., Horii, S., Yamano, T.: J. Antibiotics *28,* 298 (1975)
274. Takeda Chem. Ind. Ltd.: DOS 2 364 999, 29. 12. 1972
275. Meyer, M. C., Guttmann, D. E.: J. Pharm. Sci. *57,* 895 (1968)
276. Chignell, C. F.: Ann. Rep. Med. Chem. *1974,* 280
277. Kriegelstein, J.: Arzneimittel-Forsch. *23,* 1527 (1973)
278. Vogt, E., Lebek, G.: Schweiz. Rundschau Med. (Praxis) *62,* 441 (1973)
279. Black, J., Calesnick, B., Williams, D., Weinstein, M. J.: Antimicrobial Agents Chemother. *1963,* 138
280. Weinstein, M. J., Luedemann, G. M., Oden, E. M., Wagman, G. H.: Antimicrobial Agents Chemother. *1963,* 1
281. Gordon, R. C., Regamey, C., Kirby, W. M. M.: Antimicrobial Agents Chemother. *2,* 214 (1972)
282. Naumann, P., Auwärter, W.: Arzneimittel-Forsch. *18,* 1119 (1968)
283. Rosenkranz, H., Scheer, M., Scholtan, W.: Arzneimittel-Forsch. *26,* 1517 (1976)
284. Ramirez-Ronda, C. H., Holmes, R. H., Sanford, J. P.: J. Clin. Invest. *53,* 6, 63a (1974); Antimicrobial Agents Chemother. *7,* 239 (1975)
285. Martell, A. E., Calvin, M.: Chemistry of the metal chelate compounds, pp. 119–175, New York: Prentice Hall, Inc. *1952*
286. Foye, W. O., Lange, W. E., Swintosky, J. V., Chamberlain, R. E., Guarini, R. E.: J. Am. Pharm. Assoc. *44,* 261 (1955)
287. Yamabe, S.: J. Pharmacol. *17,* 138 (1967)
288. Irving, H., Williams, R. J. P.: Nature *162,* 746 (1948)
289. Dürckheimer, W.: unpublished
290. Mitsuhashi, S.: Transferable drug resistance factor R. Baltimore, London, Tokio: University Park Press 1971
291. Annals of the New York Academy of Sciences *182* (1971): a. Mitsuhashi, S., p. 141; b. Watanabe, T., p. 126; c. Cohen, S. N., Silver, R. P., Sharp, P. A., McCoubrey, A. E.; p. 172
292. Reinhard, E.: Pharmaz. uns. Zeit *1,* 9, 67 (1972)
293. Drug-inactivating enzymes and antibiotic resistance 2nd Intern. Symp. Antibiotic Resistance Castle of Smolenice, Czechoslovakia 1974. Mitsuhashi, S., Rosival, L., Krcinery, V. (eds.) pp. 103–192
294. Benveniste, R., Davies, J.: Antimicrobial Agents Chemother. *4,* 402 (1973)
295. Davies, J.: Ann. Rep. Med. Chemistry *7,* 217–227 (1972)

296a. Takamoto, T., Hanessian, S.: Tetrahedron Letters *1974,* 4009; b. Ogawa, T., Takamoto, T., Hanessian, S.: Tetrahedron Letters *1974,* 4013
297. Hanessian, S., Takamoto, T., Masse, R.: J. Antibiotics *28,* 835 (1975)
298. Parthier, B.: D. Pharmazie *20,* 465 (1965)
299. Hahn, F. E.: Antibiotics and Chemotherapy *17,* 29 (1971)
300. Ozaki, M., Mizushima, S., Nomura, M.: Nature *224,* 333 (1969)
301. Modolell, J., Davis, B. D.: Nature *224,* 345 (1969)
302. Biswas, D. K., Gorini, L.: Proc. Nat. Acad. Sci. *69,* 2141 (1972)
303. Miskin, R., Zamir, A.: Nature (New Biology) *238,* 78 (1972)
304. Sheran, M. I.: Europ. J. Biochem. *25,* 291 (1972)
305a. Lando, D., Cousin, M. A., Privat de Garilhe, M.: Biochemistry *12,* 4528 (1973); b. Lando, D., Cousin, M. A., Ojasoo, T., Raynaud, J. P.: Biochimie *59,* 59 (1977)
306. Stupp, H. F.: Acta Otolaryng. Suppl. *262* (1970)
307. Sheffield, P. A., Turner, Jr., J. S.: Southern Med. J. *64,* 359 (1971)
308. Wagner, W. H.: Immunität u. Infektion *3,* 241 (1975)
309. Niizato, T., Koeda, T., Tsuruoka, T., Iinouye, S., Niida, T.: J. Antibiotics *29,* 833 (1976)

Received November 2, 1978

Author Index Volumes 26–83

The volume numbers are printed in italics